Yukio Hori

Hydrodynamic Lubrication

Yukio Hori

Hydrodynamic Lubrication

With 157 Figures

Springer

Yukio Hori, Dr. Eng.
Vice President, Kanazawa Institute of Technology
7-1 Ohgigaoka, Nonoichi, Ishikawa 921-8501, Japan
Professor Emeritus, University of Tokyo

Library of Congress Control Number: 2005936105

ISBN-10 4-431-27898-2 Springer-Verlag Tokyo Berlin Heidelberg New York
ISBN-13 978-4-431-27898-6 Springer-Verlag Tokyo Berlin Heidelberg New York

Printed on acid-free paper

This English translation is based on the Japanese original:
Ryûtai Zyunkatu (Hydrodynamic Lubrication) by Yukio Hori
Published by Yokendo Ltd.
© 2002 Yukio Hori

© Springer-Verlag Tokyo 2006
Printed in Japan
This work is subject to copyright. All rights are reserved, whether the whole or part of the material is concerned, specifically the rights of translation, reprinting, reuse of illustrations, recitation, broadcasting, reproduction on microfilms or in other ways, and storage in data banks.
The use of registered names, trademarks, etc. in this publication does not imply, even in the absence of a specific statement, that such names are exempt from the relevant protective laws and regulations and therefore free for general use.

Springer is a part of Springer Science+Business Media
springeronline.com

Printing and binding: Hicom, Japan

Prefaces

To the original Japanese edition:
  Hydrodynamic lubrication occupies an important position in mechanical engineering; however, books on the subject are seldom seen in Japan. Not so many books have been published on the subject overseas either.
  This book consists of some historical and theoretical introductions (Chapters 1–4) and the results of research by myself, by myself and co-workers who were in my laboratory, and by a few of my very close colleagues (Chapters 5–9). References are given at the end of each chapter. The material has been taken from the above-mentioned research partly because of the ease in getting permission for the use of figures in published papers.
  At the manuscript stage, I received kind advice from my colleagues, approximately in the order of the contents of the book, Drs. Yoshitsugu Kimura, Masato Tanaka, Takahisa Kato, Shige-aki Kuroda, Akira Hasuike, Kyung Woong Kim, Jun'ichi Mitsui, and Satoru Kaneko. Especially, Drs. Masato Tanaka and Takahisa Kato kindly examined the whole manuscript and gave me valuable suggestions. I am very much obliged to all of them.
  Concerning publication of this book, I give sincere thanks to Mr. Kiyoshi Oikawa, the President, Mr. Nobuyuki Miura, the Director, and Mr. Kaoru Shimada, the Editor of Yokendo, Ltd..
  October 2002, Tokyo   Y.H.

To the English edition:
  On publication of the English edition, I have corrected a few errata that were found during translation and added Fig. 5.16, which was drawn after the publication of the original edition. I express my sincere thanks to the staff members of the publisher, Springer-Verlag, Tokyo.
  October 2005, Tokyo   Y.H.

Contents

1   Friction, Wear, and Lubrication 1
    1.1   Friction, Wear, and Lubrication — Tribology 1
    1.2   Various Forms of Lubrication 2
          1.2.1   Solid Friction 4
          1.2.2   Hydrodynamic Lubrication 6
    1.3   Meanings of Tribology 7
    References 8

2   Foundations of Hydrodynamic Lubrication 9
    2.1   Tower's Experiment 9
    2.2   Reynolds' Theory of Hydrodynamic Lubrication 11
          2.2.1   Interpretation of Reynolds' Equation 18
    References 22

3   Fundamentals of Journal Bearings 23
    3.1   Circular Journal Bearings 25
          3.1.1   Cross Section of a Bearing 25
          3.1.2   Shape of the Oil Film 26
          3.1.3   Bearing Length (Bearing Width) 27
          3.1.4   Boundary Conditions for the Oil Film 27
    3.2   Infinitely Long Bearings 29
          3.2.1   Oil Film Pressure 29
          3.2.2   Infinitely Long Bearing Under Sommerfeld's Condition 31
          3.2.3   Infinitely Long Bearing Under Gümbel's Condition 37
    3.3   Short Bearings 41
          3.3.1   Oil Film Pressure 41
          3.3.2   Characteristics of a Short Bearing Under Gümbel's Condition   42
    3.4   Finite Length Bearings 43
    References 46

4 Fundamentals of Thrust Bearings ... 47
4.1 Infinitely Long Plane Pad Bearings ... 48
4.1.1 Basic Formulae ... 49
4.1.2 Basic Characteristics ... 49
4.2 Finite Length Plane Pad Bearings ... 54
4.3 Sector Pad Bearings ... 55
4.3.1 Reynolds' Equation in Cylindrical Coordinates ... 55
4.3.2 Numerical Solution of a Sector Pad ... 57
4.4 Additional Topics ... 58
4.4.1 Influence of Deformation of the Pad ... 58
4.4.2 Magnetic Disk Memory Storage ... 59
References ... 60

5 Stability of a Rotating Shaft — Oil Whip ... 63
5.1 Oil Whip ... 64
5.2 Oil Whip Theory ... 67
5.2.1 Oil Film Pressure ... 68
5.2.2 Oil Film Force ... 71
5.2.3 Linearization of the Oil Film Force ... 72
5.2.4 Equations of Motion ... 75
5.2.5 Stability Limit ... 76
5.2.6 Occurrence of Oil Whip — Hysteresis ... 84
5.2.7 Coordinate Axes ... 88
5.3 Stability of Multibearing Systems ... 89
5.4 Influence of Earthquakes on Oil Whip ... 92
5.4.1 Basic Equations ... 94
5.4.2 Examples of Simulation ... 95
5.5 Limit Cycle in an Unstable Domain ... 98
5.5.1 Approximate Nonlinear Analysis of Journal Bearing Characteristics ... 98
5.5.2 Results of Analysis ... 101
5.6 Floating Bush Bearings ... 102
5.7 Three Circular Arc Bearings ... 106
5.8 Porous Bearings ... 109
5.8.1 Governing Equations ... 109
5.8.2 Stability of a Shaft System ... 110
5.9 Chaos in Rotor–Bearing Systems ... 111
5.10 Prevention of Oil Whip ... 113
References ... 114

6 Foil Bearings ... 119
6.1 Basic Equations ... 121
6.2 Finite Element Solution of the Basic Equations ... 122
6.2.1 Reynolds' Equation ... 122
6.2.2 Equation of Balance for the Foil ... 125

	6.2.3	Solution Procedure	126
6.3	Characteristics of Foil Bearings		126
	6.3.1	Single Cylinder Heads	127
	6.3.2	Double Cylinder Heads	128
	6.3.3	Comparison with Experiments	130
6.4	Additional Topics		130
	6.4.1	Magnetic Tape Memory Storage	130
	6.4.2	Foil Disk	131
References			136

7 Squeeze Film ... 137

- 7.1 Basic Equations ... 138
- 7.2 Squeeze Between Rigid Surfaces ... 141
  - 7.2.1 Squeeze Without Fluid Inertia ... 141
  - 7.2.2 Squeeze with Fluid Inertia ... 142
  - 7.2.3 Sinusoidal Squeeze Motion ... 144
- 7.3 Sinusoidal Squeeze by a Rigid Surface (Experiments) ... 145
  - 7.3.1 Mild Sinusoidal Squeeze ... 145
  - 7.3.2 Intense Sinusoidal Squeeze — Cavitation ... 146
- 7.4 Sinusoidal Squeeze with a Soft Surface ... 149
  - 7.4.1 Low-Frequency Squeeze ... 150
  - 7.4.2 High-Frequency Squeeze ... 153
  - 7.4.3 Results of Experiment and Calculation ... 154
- References ... 159

8 Heat Generation and Temperature Rise ... 161

- 8.1 Basic Equations for Thermohydrodynamic Lubrication ... 162
- 8.2 Generalized Reynolds' Equation ... 163
  - 8.2.1 Balance of Forces ... 163
  - 8.2.2 Flow Velocity ... 164
  - 8.2.3 Continuity Equation ... 164
  - 8.2.4 Generalized Reynolds' Equation ... 165
- 8.3 Energy Equation ... 166
  - 8.3.1 General Energy Equation ... 166
  - 8.3.2 Energy Equation ... 168
  - 8.3.3 Transformation of the Energy Equation ... 170
- 8.4 Temperature Distribution in Bearings ... 171
- 8.5 Temperature Analyses of Tilting Pad Thrust Bearings — Sector Pads 172
  - 8.5.1 Basic Equations ... 173
  - 8.5.2 Boundary Conditions ... 175
  - 8.5.3 Numerical Analyses ... 175
  - 8.5.4 Examples of Three-Dimensional Analyses of Temperature Distribution ... 177
  - 8.5.5 Comparisons of Three-Dimensional, Two-Dimensional, and Isoviscous Analyses ... 178

		8.5.6	Analysis Considering Inertia Forces	180
		8.5.7	Comparison of Calculated Results and Experiments	184
	8.6	Temperature Analyses of Circular Journal Bearings		185
		8.6.1	Basic Equations	187
		8.6.2	Boundary Conditions	187
		8.6.3	Comparison of Calculated Results and Experiments	189
	References			193
9	**Turbulent Lubrication**			197
	9.1	Time-Average Equation of Motion and the Reynolds' Stress		198
	9.2	Turbulent Flow Model		201
		9.2.1	Mixing Length Model	201
		9.2.2	k-ε Model	203
	9.3	Turbulent Lubrication Theory Using the Mixing Length Model		204
		9.3.1	Modified Mixing Length	204
		9.3.2	Turbulent Velocity Distribution Between Two Surfaces	206
		9.3.3	Turbulent Reynolds' Equation	208
		9.3.4	Turbulent Coefficients of Fluid Film Seals	209
	9.4	Comparison of Analyses Using the Mixing Length Model with Experiments		211
		9.4.1	Turbulent Static Characteristics of Fluid Film Seals	211
		9.4.2	Turbulent Dynamic Characteristics of Fluid Film Seals	213
	9.5	Turbulent Lubrication Theory Using the k-ε Model		214
		9.5.1	Application of the k-ε Model to an Oil Film	215
		9.5.2	Turbulent Reynolds' Equation	216
	9.6	Comparison of Analyses Using the k-ε Model with Experiments		218
	9.7	Reduction of Friction in a Turbulent Bearing by Toms' Effect		222
	9.8	Taylor Vortices in a Journal Bearing		224
	References			226
Index				229

The references at the end of each chapter are listed as a rule in chronological order.

Symbols

The meanings of symbols are as follows, unless otherwise stated.

A	: area		solid parts
b	: bearing width (length)	k_x, k_z	: reciprocal of turbulent coefficient
B	: pad width		
B_x, B_z	: nondimensional pressure gradient	l	: mixing length
		l_m	: modified mixing length
B_p	: bearing parameter	L	: bearing length (width)
c	: radial clearance ($= R_b - R_j$)	L_s	: frictional loss
c_p	: specific heat at constat pressure	m	: inclination of pad ($= h_1/h_2$)
c_v	: specific heat at constat volume	M	: frictional moment
D	: diameter	N	: rotational speed of shaft
e	: eccentricity ($= \overline{O_b O_j}$)	p	: pressure, film pressure
f	: coefficient of friction (Section 1.2.1)	p_m	: bearing pressure ($= P/(2DL)$)
		P	: oil film force
f	: frequency of squeeze motion (Chapter 7)	P_0	: oil film force (resultant)
		P_1	: bearing load
F	: force	$\overline{P}$	: nondimensional load capacity
G_x, G_z	: turbulent coefficient	q	: flow rate of oil
G_c	: term of centrifugal force	Q	: heat flux
h	: film thickness	Q_s	: generated heat
h	: enthalpy (Section 8.3.3)	r, θ	: polar coordinates
h_c	: coefficient of heat transfer	R	: radius
H	: indentation hardness (Section 1.2.1)	Re	: bearing Reynolds' number ($= Uc/\nu$)
H	: mean stress function (Section 7.4.1)	R_h	: local Reynolds' number ($= Uh/\nu$)
J	: functional	R_t	: turbulent Reynolds' number ($= k^2/\varepsilon\nu$)
k	: turbulent energy		
k_o, k_s	: thermal conductivity of oil and	r, θ, z	: cylindrical coordinates

XII   Symbols

S	: Sommerfeld number $(= (R/c)^2 \mu N / p_m)$	κ	: eccentricity ratio $(= e/c)$
		κ_k	: Karman's constant
t	: time	λ	: secondary coefficient of viscosity
t	: thickness of pad (Section 8.5)	μ	: coefficient of viscosity
T	: tension (Chapter 6)	ν	: coefficient of kinetic viscosity $(= \mu/\rho)$
T	: temperature		
T_2	: temperature of pad (Section 8.5)	ν	: Poisson's ratio (Section 7.4.1)
		Π	: functional
u, v, w	: components of fluid velocity	ρ	: density
u^*	: frictional velocity $(= \sqrt{\tau_w/\rho})$	σ	: normal stress
u^+	$:= u/u^*$	σ_y	: yield stress (Section 1.2.1)
U, V, W	: surface velocity	τ	: shear stress
U	: internal energy (Section 8.3)	τ_w	: surface shear stress
V	: generalized velocity (Section 8.3)	τ^+	$:= \tau/\tau_w$
		ϕ	: angle
x, y, z	: rectangular coordinate	Φ	: permeability (Section 5.9)
y^+	$:= u^* y/\nu$	Φ	: dissipation energy (Chapter 8)
X, Y, Z	: rectangular coordinate axes	φ	: angle (Sommerfeld transform)
α	: coefficient of expansion	ψ	: angle
α_v	: coefficient of cubic expansion	ω	: angular velocity of rotation
β	: wrap angle (Chapter 6)	Ω	: angular velocity of whirling
β	: viscosity index (Chapter 8)		
δ_{ij}	: Kronecker's delta	**Suffix**	
ϵ	: strain	i, j, k	: 1,2,3
ϵ	: small quantity (Sections 3.4, 6.2.3)	a	: ambient
		b	: bearing
ε	: turbulent dissipation	j	: journal
η	: mixing coefficient	s	: solid part
θ	: attitude angle	t	: turbulent
		$(\tilde{\ })$	: non-dimensional

1

Friction, Wear, and Lubrication

1.1 Friction, Wear, and Lubrication — Tribology

Since the dawn of history, human activities have always been closely related to **friction**, the resistance to sliding. It is thanks to friction that one can stand and walk on the ground, one can wear clothes, one can make fire by rubbing two sticks together, or one can even start and stop a car. In these cases friction is very useful for human beings. In many other cases, however, human activities have been very much hampered by friction since ancient times. How to diminish friction is one of the most basic technological problems.

For example, when a heavy object is moved on a floor, oil or water are used between the bottom of the object and the floor to reduce the high level of friction between them. This is one of the technologies about which human beings worldwide have known from the most ancient time. Rollers have also been used for the same purpose for many thousands of years. Human beings have made every effort to overcome friction in many fields since the very beginning of history.

The situation has not changed very much, even after the advent of machine-based civilizations. There are many pairs of machine parts that are in relative motion such as journals and bearings and the teeth of gears. In such cases, friction always exists between the two sliding surfaces. As a result, not only **energy loss** and **wear** but also **seizure** may take place. It is the basis of machine technology to prevent these phenomena by supplying a suitable substance such as oil.

In some linkage mechanisms, some of the links cannot move no matter how big a force is applied because the friction in the joints also increases in proportion to the applied force. This phenomenon is called self-locking.

Reducing friction, wear, and the occurence of seizure by providing a suitable substance such as oil between two surfaces in relative motion is called **lubrication**, and the substance used for this purpose is called a **lubricant**. Beside oils, which are typical lubricants, other liquids such as water or emulsion or even some gases are used as lubricants. Some solids are also used as a lubricant, mostly in the form of film coatings.

Generally speaking, suitable lubrication can reduce friction and hence energy consumption. Furthermore, suitable lubrication can reduce wear and prevent seizure and hence can extend the life of a machine and thus save natural resources. Therefore suitable lubrication is useful for **saving energy** and **resources**. This is very important in the light of the finiteness of natural resources on the earth. In fact, the effect of lubrication is usually far more remarkable in reducing wear than in reducing friction.

The word **tribology** has often been heard in recent years. The word was first used in 1966 in the Jost Report of the British Department of Education and Science. To emphasize the importance of science and technology concerning friction, wear, and lubrication, these classical subjects were unified in the report under a new name, "tribology." Its definition is "the science and technology of interacting surfaces in relative motion and practices related thereto." The word tribology has permeated not only among specialists but also among the public and is now used worldwide. Although the history of tribology is very old in terms of technology, it is comparatively new as a science. The importance of tribology will grow increasingly in the future from the viewpoint of the conservation of energy and resources. The importance of tribology is also recognized in new fields such as memory device technology, space engineering and bioengineering, and new words such as microtribology, space tribology, and biotribology have been coined. The word tribology is based on the Greek word *tribos*, which means rubbing.

It is said that the background to the birth of the concept of tribology was as follows. When the Jost Committee of the British Department of Education and Science studied the economic loss as a result of friction, wear, and seizure in Britain in the 1960s, it came up with a figure of more than £515 million per year. This was a signigicant amount compared with the national budget of Britain at that time. Therefore, it was considered very useful if the loss could be reduced by research and development of related technologies. It was planned to promote the research program on friction, wear, and lubrication efficiently, and to do this, the importance of the research had to be signaled to the public. It was thought that the best way for this was to unify these classical subjects as one concept and to give it a new name — tribology.

1.2 Various Forms of Lubrication

Although the main subject of this book is hydrodynamic lubrication, it is worthwhile to initially consider the various forms of lubrication.

It is known that the coefficient of friction of a journal bearing changes with operating conditions as shown in Fig. 1.1. The vertical axis indicates the coefficient of friction $f = F/P$ and the horizontal axis the bearing number $\mu U/P$, where F = frictional force, P = journal load, μ = coefficient of viscosity, and U = circumferential velocity of the journal (the part of a shaft supported by a bearing). The coefficient of friction f has a minimum point as shown in the figure. The value of f at the minimum point is very small, usually of the order of 0.001. For larger values of bearing number $\mu U/P$, the coefficient of friction f increases along a straight line passing through the origin. The rate of increase is small. With a decrease in the bearing number from

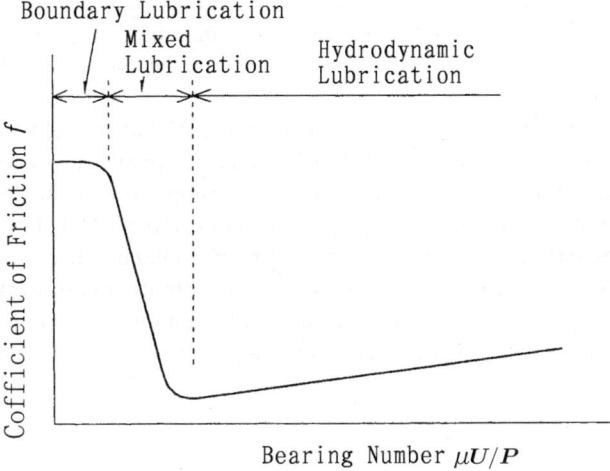

Fig. 1.1. Stribeck diagram

the point of minimum coefficient of friction, in contrast, the frictional coefficient increases rapidly, but does not exceed a certain fixed value. Since the diagram is based on the careful, extensive experiments (1902) carried out by Richard Stribeck (1861–1950) of Germany, it is called the **Stribeck diagram**. The diagram exhibits clearly the features of the frictional coefficient of a journal bearing.

The reason why the curve in the Stribeck diagram takes such a form is as follows. First, consider the region where the bearing number is sufficiently large (the region on the right of the minimum point, or the region where, for example, the circumferential speed is sufficiently high). In this region, the frictional coefficient f increases at a very low rate toward larger values of the bearing number, its value being of the order of 0.001. The reason for this is that a sufficiently thick oil film is formed between the two surfaces in relative motion, and the two surfaces do not contact each other directly. The frictional force in this case is attributable to the viscosity of oil and is proportional to the shear rate of the oil film. The bearing load is supported by the pressure produced in the oil film. Since the two surfaces do not contact directly, wear hardly takes place. This is an ideal state of lubrication and is called **hydrodynamic lubrication**.

If the bearing number is lowered (or if, say, the circumferential speed is lowered) in this state, the frictional resistance will decrease gently toward the minimum point. The oil film becomes gradually thinner and at a certain bearing number, the oil film becomes so thin that the minute projections of the two surfaces finally begin to collide with each other. Therefore the frictional resistance no longer decreases but increases rapidly, producing a minimum point. With further decrease in the bearing number, collision of the projections of the two surfaces becomes severer and the frictional coefficient will increase further. The situation at this stage is a mixture of the hydrodynamic lubrication described above and boundary lubrication which will be explained later and is called **mixed lubrication**. Since all surfaces in practice, however smooth they may be, have minute asperities, the above-described process is always followed.

With further decrease in the bearing number, the frictional coefficient will finally reach a magnitude of the order of 0.1. In this state, as a result of contact between the asperities of the two surfaces, the load is mostly supported by the solid contact of asperities and the hydrodynamic role of the oil film is largely lost. The state is called **boundary lubrication**. A frictional surface under boundary lubrication conditions is schematically shown in Fig. 1.2. The solid surface in the figure is mostly covered with a thin adsorption layer of oil molecules, although the layer is destroyed where the projections are rubbing severely. Seizure may take place at such locations.

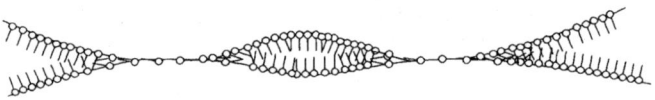

Fig. 1.2. Oil film under boundary lubrication conditions. ↑denotes an oil molecule. Polar end-groups on the oil molecules bond to the surfaces

In all the cases described above, some lubricating oil exists on the sliding surfaces. Now consider the friction of surfaces which are made as clean as possible in an ordinary sense by, for example, wiping off the lubricant with alcohol. The friction in this case is called **dry friction**. The coefficients of friction of various materials listed in data books are usually those of dry friction. Their magnitudes are usually in the range of 0.1 – 1, although they exhibit considerable variation. In this case, the surfaces are actually covered with a small amount of foreign substances such as oxide film. If the foreign substances are removed from the surfaces by, for example, heating them in a vacuum, the friction may become close to the ideal dry friction. In this case, however, the frictional behavior of materials is usually quite different from those in ordinary conditions; the surfaces may often adhere to each other. It is difficult to distinguish ordinary dry friction from ideal dry friction.

1.2.1 Solid Friction

The friction between two solid surfaces in the case of dry friction or boundary lubrication is called **solid friction**. For solid friction, the following law is well known.

> The frictional force F is proportional to perpendicular load P, and is independent of the apparent contact area and the sliding speed.

In mathematical form, it can be written as follows:

$$F = fP \tag{1.1}$$

where f is the coefficient of friction.

It is said that this law was first discovered by Leonardo da Vinci (1452–1519) of Italy in the fifteenth century, and was later rediscovered by G. Amonton (1663–1705) and C. A. Coulomb (1736–1806) of France in the eighteenth century independently. Today, the law is called Amonton's law or Coulomb's law.

An explanation for the fact that the frictional force is proportional to the perpendicular load and is not related to the apparent contact area is as follows. Since all solid surfaces have small asperities, when two solid surfaces are in contact, they actually contact each other only through small projections of the surfaces (See Fig. 1.2). The area of true contact in this case, which is called the **true contact area**, is usually very small. In view of this situation, the so-called frictional force will be the shearing force necessary to overcome the adhesion of the material over the true contact area. Thus, if the true contact area is denoted by A_r and the shearing strength of the adhesion per unit area of true contact by s_m, the frictional force F will be $F = s_m A_r$.

Now, the true contact area A_r can be estimated as follows. Since the true contact area is very small, the contact stress in the true contact area is very high, and so the material yields there. Therefore, if the load is P and the yield stress of the material is σ_y, then the true contact area is given by $A_r = P/\sigma_y$. Thus, the true contact area is proportional to the load.

Combining the above considerations will give the frictional force F for dry friction as follows.

$$F = s_m A_r = s_m (P/\sigma_y)$$

In other words, the frictional force is not related to the apparent area of contact but is proportional to the load. Now, from the above equation, the frictional force F can be written as follows with the coefficient of friction f:

$$F = fP \quad \text{where} \quad f = s_m/\sigma_y \tag{1.2}$$

The coefficient of friction is thus given by the ratio of the shearing strength of adhesion to the yield stress of the material.

For boundary lubrication, the true contact area A_r can be divided into the area of direct contact αA_r and that of contact through a coating film $(1 - \alpha)A_r$. If the shear strength of the contact area through the film per unit area is denoted by s_l, the frictional force F will be given as follows:

$$F = \{s_m \alpha + s_l(1 - \alpha)\}A_r \tag{1.3}$$

or

$$F = fP \quad \text{where} \quad f = \{s_m \alpha + s_l(1 - \alpha)\}/\sigma_y \tag{1.4}$$

6   1 Friction, Wear, and Lubrication

The frictional force is proportional to the load in this case also. It is clear in the above argument that only the true contact area is related to the friction, not the apparent contact area.

Such a view is called the **adhesion theory** of the friction. The concept of the true contact area was first introduced by Ragnar Holm [1] in relation to electric contacts.

1.2.2 Hydrodynamic Lubrication

Hydrodynamic lubrication, which is the subject of this book, is an ideal state of lubrication in that friction and wear hardly occur. Figure 1.3 shows three typical examples in which hydrodynamic lubrication is important. These are a journal bearing, an air-floating slider in a magnetic disk memory device, and an animal joint.

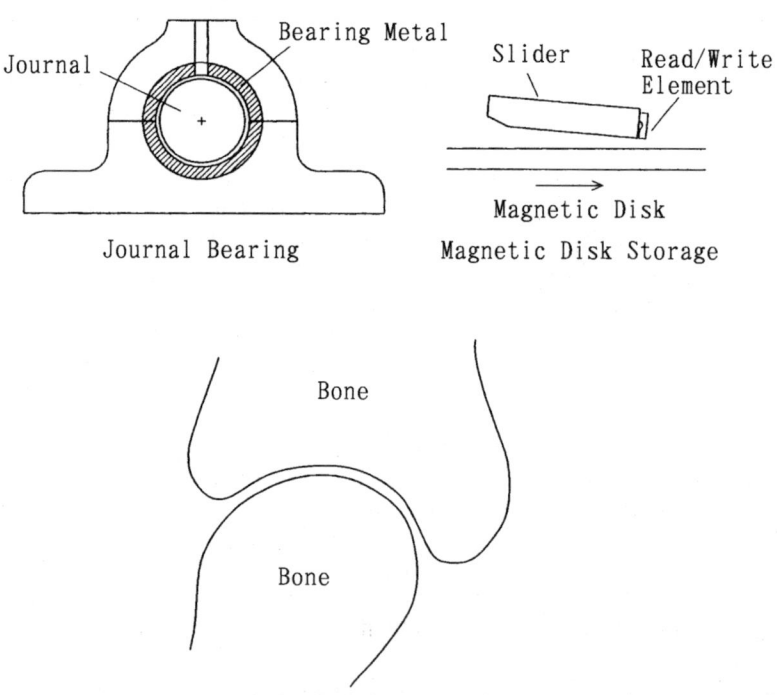

Fig. 1.3. Examples of hydrodynamic lubrication

The first is a journal bearing. In this case, since the journal center is slightly displaced from the bearing center in a diagonally downward direction due to the bearing load, the clearance between the journal and the bearing metal varies in the circumferential direction. In the region where the clearance becomes smaller in the direction of journal rotation, the oil film forms a wedge and pressure is generated in it due to the journal rotation. This is called the **wedge effect** of an oil film. The bearing load is supported by the oil film pressure and the journal floats on the oil film. Therefore, the frictional resistance is very small. Journal bearings range in size from the very small, such as those supporting rotating grinders for dentistry, to the very large, such as those supporting steam turbines, generators, and hydraulic turbines, for example.

The second is a magnetic disk memory device for computers. A slider with a magnetic head, or a read/write element, at its trailing edge floats on a very thin air film on the surface of a rotating magnetic disk. In this case, the ambient air is drawn into the wedge-shaped clearance between the disk and the slider, forming an air film wedge and generating pressure in it. The magnetic head is supported in this way without contacting the disk. The smaller the clearance between the magnetic head and the magnetic disk, the higher the recording density can be, but the slider and the magnetic head are not allowed to contact the disk from the viewpoint of crash prevention. The magnitude of the clearance in recent devices is of the order of 10–20 nm.

The third is a skeletal joint. A lubricating film of synovial fluid exists between the cartilage-covered bone surfaces that constitute the joint. As the synovial fluid is squeezed out by the weight, a pressure is produced in the synovia, and this pressure, although transient, supports the weight. This is called the **squeeze effect** of a lubricating film. During the loading period, the lubricating film in the joint becomes thinner, whereas during the unloading period, the thickness is recovered. Reynolds pointed out in his first paper (1886) that the squeeze effect is of fundamental importance in an animal joint. He had a really keen insight.

1.3 Meanings of Tribology

Tribology is often likened to an unsung hero. In other words, tribology often plays only an inconspicuous role in machine technology. In reality, however, tribology supports machine technology very fundamentally. It often plays an important role that determines the fate of a machine. A competent engineer, even in fields other than tribology, recognizes this. Some comments of such specialists are given below.

"In the case of steam turbines, airfoil theories may attract much attention, but I believe the journal bearings supporting the turbines are also very important. If seizure occurs in a bearing, the turbine will stop. In other words, a bearing determines the fate of the whole machine. In this connection, bearing technology is very important in a completely different way from airfoil theory, which may improve turbine efficiency by 2% or 3% by changing the airfoil of turbine blades."

"In the case of a magnetic disk memory device, hydrodynamic lubrication develops between a slider and the magnetic disk, with the surrounding air as lubricant. If seizure (or a crash) takes place there, it's all over. All records will be lost. All high technologies in the other parts of the system will be useless. The importance of tribology cannot be overstated."

"Present-day robots are not yet so developed that they need tribology. It is still a question of whether they move as planned or not. Once this question is settled, the problem will arise of how they move, and then tribology will become necessary."

"The antenna didn't open because of seizure. The satellite doesn't work. It's a loss of 10 billion yen. All because of tribology ... "

References

1. R. Holm, "Electric Contacts", *H. Gebers Förlag*, Stockholm, 1946.
2. F.P. Bowden and D. Tabor, "The Friction and Lubrication of Solids - Part I", *Oxford U.P.*, Oxford, 1950.
3. Norimune Soda, "Friction and Lubrication" (in Japanese), *Iwanami Zensho Series, Iwanami Shoten*, Tokyo, 1954.
4. D. Dowson, "History of Tribology", *Longman*, London, 1979.
5. Yoshitugu Kimura and Heihachiro Okabe, "Introduction to Tribology" (in Japanese), *Yokendo Ltd.*, Tokyo, 1982.

2

Foundations of Hydrodynamic Lubrication

The essence of hydrodynamic lubrication was first clarified experimentally by British railroad engineer Beauchamp Tower (1845–1904) in 1883 [1][2]. Based on Tower's experiments, Osborn Reynolds (1842–1912), the physicist, formulated a theory of lubrication in 1886 [3]. Since then, Reynolds' theory has been the foundation of the theory of hydrodynamic lubrication. Recently developed theories of elastohydrodynamic lubrication, thermohydrodynamic lubrication, turbulent hydrodynamic lubrication, and others are regarded as extensions of Reynolds' theory. The pioneering works of Tower and Reynolds are reflections of Britain's advanced technology at that time.

In this chapter, Tower's experiment will be explained first and then Reynolds' theory will be derived.

2.1 Tower's Experiment

Figure 2.1, which is a simplification of a drawing from Tower's famous paper of 1883 [1], shows the main part of Tower's friction test rig for a bearing used in rolling stock. The bearing is a partial bearing, and bearing bush A covers the upper half of the journal. A load (weight of the vehicle) acts on the journal from above through bearing cap B and bearing bush A.

The lower part of the journal is immersed in lubricating oil, and the oil adhering to the journal surface is pulled up by rotation of the journal and is supplied to the bearing clearance. Such a method of lubrication is called oil bath lubrication. The frictional resistance (frictional moment) of the bearing can be obtained by measuring the frictional moment acting on the bearing cap.

Using this test rig, Tower found the frictional characteristics of a journal bearing with oil bath lubrication to be as follows.

1. Frictional resistance is nearly constant, regardless of the bearing load.
2. The frictional coefficient is very small (usually of the order of 1/1000).
3. Frictional resistance increases with sliding speed.

2 Foundations of Hydrodynamic Lubrication

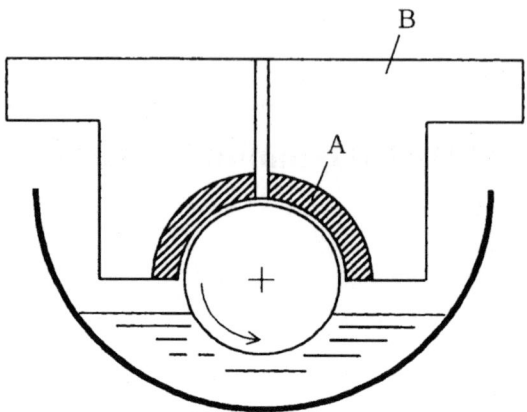

Fig. 2.1. Tower's test rig. A, bearing bush; B, bearing cap

4. Frictional resistance decreases with a rise in temperature.

Furthermore, he pointed out that the friction in this case followed the laws of "liquid friction" much more closely than those of the solid friction (Coulomb friction). Moreover, he reported the following interesting observations:

> A very interesting discovery was made when the oil-bath experiments were on the point of completion. ··· While the brass was out, the opportunity was taken to drill a 1/2-inch hole for an ordinary lubricator through the cast-iron cap and the brass. On the machine being put together again and started with the oil in the bath, oil was observed to rise in the hole which had been drilled for the lubricator. The oil flowing over the top of the cap made a mess, and an attempt was made to plug up the hole, first with a cork and then with a wooden plug. When the machine was started the plug was slowly forced out by the oil in a way which showed that it was acted on by a considerable pressure. A pressure-gauge was screwed into the hole, and on the machine being started the pressure, as indicated by the gauge, gradually rose to above 200 lbs. per square inch. The gauge was only graduated up to 200 lbs., and the pointer went beyond the highest graduation. The mean load on the horizontal section of the journal was only 100 lbs. per square inch. This experiment showed conclusively that the brass was actually floating on a film of oil, subject to a pressure due to the load. The pressure in the middle of the brass was thus more than double the mean pressure. No doubt if there had been a number of pressure-gauges connected to various parts of the brass, they would have shown that the pressure was highest in the middle, and diminished to nothing towards the edges of the brass.

This is exactly what we call hydrodynamic lubrication today. In the last part, Tower even referred to the oil pressure distribution, which means that he had had a keen insight into the problem. Actually, in his second paper [2], Tower reported that the beautiful pressure distribution was observed as was expected.

In the case of usual lubrication methods (lubrication other than oil bath lubrication), Tower reported that the measured value of friction was often unstable. Probably, the quantity of oil was inadequate and a perfect oil film was not formed. In the case of the oil bath lubrication, in contrast, a sufficient amount of oil was presumably supplied.

2.2 Reynolds' Theory of Hydrodynamic Lubrication

Reynolds was interested in Tower's experiments and studied them theoretically. In the introductory part of his famous paper of 1886 [3] on hydrodynamic lubrication, Reynolds wrote:

> Lubrication, or the action of oils and other viscous fluids to diminish friction and wear between solid surfaces, does not appear to have hitherto formed a subject for theoretical treatment. Such treatment may have been prevented by the obscurity of the physical actions involved, which belong to a class as yet but little known, namely, the boundary or surface actions of fluids; but the absence of such treatment has also been owing to the want of any general laws discovered by experiment.
>
> The subject is of such fundamental importance in practical mechanics, and the opportunities for observation are so frequent, that it may well be a matter of surprise that any general laws should have for so long escaped detection.
> ....
>
> On reading Mr. Tower's report it occurred to the author it is possible that in the case of the oil bath the film of oil might be sufficiently thick for the unknown boundary actions to disappear, in which case the results would be deducible from the equations of hydrodynamics.

Phenomena related to lubrication are, generally speaking, very complicated because of, for example, the complexity of interface phenomena and their theoretical treatments. In the case of oil bath lubrication, however, Reynolds assumed that the oil film was so thick that the theory of hydrodynamics could be applied.

The outline of Reynolds theory is now described. The fluid film between two solid surfaces shown in Fig. 2.2 is considered. Reynolds' equation is an equation to obtain the pressure generated in a fluid film when two such surfaces undergo relative motion. However, the fluid film must be sufficiently thick so that it can be alalyzed by hydrodynamics, and at the same time it must be sufficiently thin so that Reynolds' assumptions described below will hold. For simplicity, the lower surface is assumed to be a plane.

The axes of rectangular coordinates x, y, and z are taken as shown in the figure. The x and z axes are on the lower surface, and the y axis is perpendicular to it. The velocity of the fluid in the directions x, y, and z are denoted by u, v, and w,

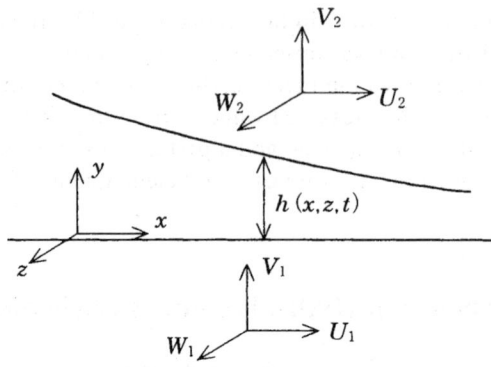

Fig. 2.2. Fluid film between two solid surfaces

respectively, and the velocity of the lower surface is similarly described by U_1, V_1, and W_1 and that of the upper surface by U_2, V_2, and W_2. In many practical cases, the lower surface and the upper surface perform a straight translational motion relative to each other. In this case, if the x axis is in the translational direction, then we have $W_1 = W_2 = 0$ and so the equations can be simplified.

Let the gap between the two surfaces, or the thickness of the liquid film, be denoted by $h(x, z, t)$, with t being time. Let the coefficient of viscosity of the fluid be μ.

a. Reynolds' Assumptions

In deriving Reynolds' equation, the following assumptions are made after Reynolds.

1. The flow is laminar.
2. The gravity and inertia forces acting on the fluid can be ignored compared with the viscous force.
3. Compressibility of the fluid is negligible.
4. The fluid is Newtonian and the coefficient of viscosity is constant.
5. Fluid pressure does not change across the film thickness.
6. The rate of change of the velocity u and w in the x direction and z direction is negligible compared with the rate of change in the y direction.
7. There is no slip between the fluid and the solid surface.

b. Balance of Forces

The balance of forces acting on a small volume element in the fluid is considered as shown in Fig. 2.3.

Let us examine the balance in the x direction first. Neglecting the gravity and inertia forces (assumption 2), we obtain the following equation:

2.2 Reynolds' Theory of Hydrodynamic Lubrication

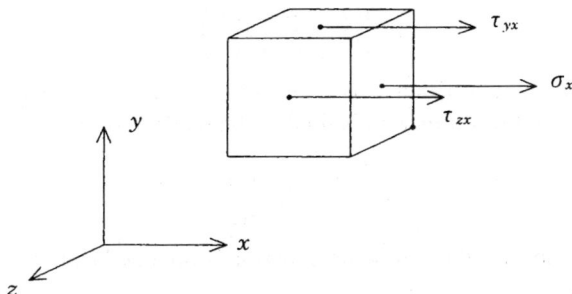

Fig. 2.3. A small element of fluid

$$\left(\sigma_x + \frac{\partial \sigma_x}{\partial x}dx\right)dydz + \left(\tau_{yx} + \frac{\partial \tau_{yx}}{\partial y}dy\right)dxdz$$
$$+ \left(\tau_{zx} + \frac{\partial \tau_{zx}}{\partial z}dz\right)dxdy - \sigma_x dydz - \tau_{yx}dxdz - \tau_{zx}dxdy = 0 \quad (2.1)$$

where σ_x is the normal stress acting on the plane normal to the x axis and τ_{yx} and τ_{zx} are the shear stresses acting on the plane normal to the y axis and z axis, respectively, in the direction of the x axis.

Equation 2.1 can be rearranged as follows:

$$\frac{\partial \sigma_x}{\partial x} + \frac{\partial \tau_{yx}}{\partial y} + \frac{\partial \tau_{zx}}{\partial z} = 0 \quad (2.2)$$

Let the fluid pressure be p. Then $p = -\sigma_x$ and the above equation can be written as follows:

$$\frac{\partial p}{\partial x} = \frac{\partial \tau_{yx}}{\partial y} + \frac{\partial \tau_{zx}}{\partial z} \quad (2.3)$$

Since a laminar flow of Newtonian fluid is considered here (assumptions 1 and 4), we have the following relations.

$$\tau_{yx} = \mu \frac{\partial u}{\partial y} \qquad \tau_{zx} = \mu \frac{\partial u}{\partial z} \quad (2.4)$$

where μ is the coefficient of viscosity. Then Eq. 2.3 can be written as follows:

$$\frac{\partial p}{\partial x} = \frac{\partial}{\partial y}\left(\mu \frac{\partial u}{\partial y}\right) + \frac{\partial}{\partial z}\left(\mu \frac{\partial u}{\partial z}\right) \quad (2.5)$$

On the assumption that the rate of change of the flow velocity u in the z direction is sufficiently small compared with that in the y direction (assumption 6), the second term of the right-hand side of the above equation can be disregarded compared with the first term, giving:

$$\frac{\partial p}{\partial x} = \frac{\partial}{\partial y}\left(\mu \frac{\partial u}{\partial y}\right) \qquad (2.6)$$

On the further assumption that μ is constant (assumption 4), the equation of the balance of forces in the x direction is finally obtained as follows:

$$\frac{\partial p}{\partial x} = \mu \frac{\partial^2 u}{\partial y^2} \qquad (2.7)$$

In exactly the same way, the following equation is obtained from the balance in the z direction:

$$\frac{\partial p}{\partial z} = \mu \frac{\partial^2 w}{\partial y^2} \qquad (2.8)$$

c. Flow Velocity

Integrating Eqs. 2.7 and 2.8 twice gives the flow velocity u and w, respectively. The boundary conditions for the velocities are, from the assumption that there is no slip between the fluid and the solid surface (assumption 7), as follows:

$$\left.\begin{array}{l} u = U_1, \; w = W_1 \; \text{at} \; y = 0 \\ u = U_2, \; w = W_2 \; \text{at} \; y = h \end{array}\right\} \qquad (2.9)$$

Then the fluid velocities will be as follows:

$$u = -\frac{1}{2\mu}\frac{\partial p}{\partial x}y(h-y) + \left[\left(1-\frac{y}{h}\right)U_1 + \frac{y}{h}U_2\right] \qquad (2.10)$$

$$w = -\frac{1}{2\mu}\frac{\partial p}{\partial z}y(h-y) + \left[\left(1-\frac{y}{h}\right)W_1 + \frac{y}{h}W_2\right] \qquad (2.11)$$

where, in the calculations, it is assumed that the pressure p is constant in the y direction (assumption 5).

In Eq. 2.10 for the flow velocity u, the latter half of the right-hand side (in brackets) shows the fluid velocity due to the movement of the solid surface in the x direction. It changes linearly as shown in Fig. 2.4a (it is assumed that $U_2 = 0$). This is called **shear flow** or **Couette flow**. The former half of the right-hand side shows the flow velocity due to the pressure gradient. It is proportional to the pressure and changes parabolically across the film thickness as shown in Fig. 2.4b. This is called **pressure flow** or **Poiseuille flow**. The flow velocity in a general case is the sum of the two. Figure 2.4c shows such an example in which a flow in the reverse direction to the shearing direction occurs due to the pressure gradient at the left end and the shear flow is accelerated by the negative pressure gradient at the right end. At the point of maximum pressure, we have the relation $dp/dx = 0$, and therefore the flow at that point consists of the shear flow only.

The same can be said of Eq. 2.11 for the flow velocity w.

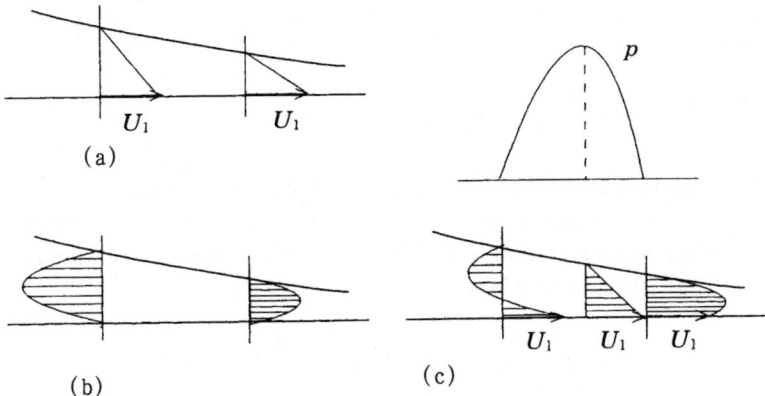

Fig. 2.4a-c. Flow velocity distribution. **a** shear flow, **b** pressure flow, **c** summation of shear flow and pressure flow

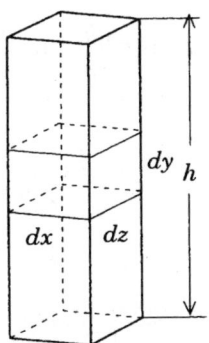

Fig. 2.5. Pillar-shaped element

d. Continuity Equation

The continuity equation for a small volume element in an incompressible fluid (assumption 3) can be written as follows:

$$\frac{\partial u}{\partial x} + \frac{\partial v}{\partial y} + \frac{\partial w}{\partial z} = 0 \tag{2.12}$$

Integrating this equation along the pillar-shaped element ($dx \times dz \times h$) in Fig. 2.5 in the film thickness direction from $y = 0$ to $y = h$ gives:

$$\int_0^h \frac{\partial u}{\partial x} dy + \int_0^h \frac{\partial w}{\partial z} dy + \left[v \right]_0^h = 0 \tag{2.13}$$

This equation can be rewritten as follows by changing the order of integration and differentiation:

$$\frac{\partial}{\partial x}\int_0^h u\,dy + \frac{\partial}{\partial z}\int_0^h w\,dy - (u)_{y=h}\frac{\partial h}{\partial x} - (w)_{y=h}\frac{\partial h}{\partial z} + [v]_0^h = 0 \qquad (2.14)$$

This is the continuity equation for a pillar-shaped element between the solid surfaces shown in Fig. 2.6. In deriving Eq. 2.14, the following mathematical formula must be used because the integration limit h in Eq. 2.13 is a function of x and z (only a formula concerning x is given here).

$$\frac{\partial}{\partial x}\int_{a(x)}^{b(x)} f(x,y,z)\,dy = \int_{a(x)}^{b(x)} \frac{\partial}{\partial x} f(x,y,z)\,dy$$
$$+ f(x,b(x),z)\frac{\partial b(x)}{\partial x} - f(x,a(x),z)\frac{\partial a(x)}{\partial x}$$

Equation 2.14 can be written as follows in terms of the surface velocities from the boundary conditions (assumption 7):

$$\frac{\partial}{\partial x}\int_0^h u\,dy + \frac{\partial}{\partial z}\int_0^h w\,dy - U_2\frac{\partial h}{\partial x} - W_2\frac{\partial h}{\partial z} + (V_2 - V_1) = 0 \qquad (2.15)$$

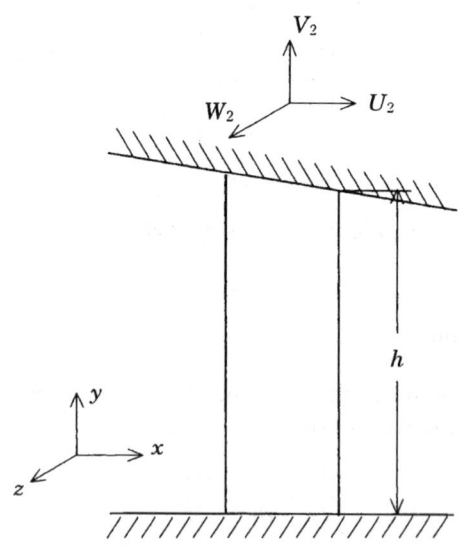

Fig. 2.6. An inclined plate in motion

e. Derivation of Reynolds' Equation

Reynolds' equation is essentially a continuity equation. Substituting the flow velocities u and v given by Eqs. 2.10 and 2.11, respectively, into continuity equation Eq. 2.15 yields Reynolds' equation.

Before doing this, the integrals (flow rates) in Eq. 2.15 are calculated:

$$\int_0^h u\, dy = -\frac{h^3}{12\mu}\frac{\partial p}{\partial x} + \frac{h}{2}(U_1 + U_2) \tag{2.16}$$

$$\int_0^h w\, dy = -\frac{h^3}{12\mu}\frac{\partial p}{\partial z} + \frac{h}{2}(W_1 + W_2) \tag{2.17}$$

where μ and p are assumed to be constant in the y direction (assumption 4, assumption 5). Substituting these two integrals into Eq. 2.15 gives the following equation:

$$\frac{\partial}{\partial x}\left(\frac{h^3}{\mu}\frac{\partial p}{\partial x}\right) + \frac{\partial}{\partial z}\left(\frac{h^3}{\mu}\frac{\partial p}{\partial z}\right) = 6\left[(U_1 - U_2)\frac{\partial h}{\partial x} + h\frac{\partial}{\partial x}(U_1 + U_2)\right.$$
$$\left. + (W_1 - W_2)\frac{\partial h}{\partial z} + h\frac{\partial}{\partial z}(W_1 + W_2) + 2(V_2 + V_1)\right] \tag{2.18}$$

In many practical cases, the x axis can be taken as the direction of the relative motion of the two surfaces, in which case we have $V_1 = W_1 = W_2 = 0$. If the coefficient of viscosity μ is constant (assumption 4), the above equation can be simplified as follows:

$$\frac{\partial}{\partial x}\left(h^3\frac{\partial p}{\partial x}\right) + \frac{\partial}{\partial z}\left(h^3\frac{\partial p}{\partial z}\right) = 6\mu\left[(U_1 - U_2)\frac{\partial h}{\partial x} + h\frac{\partial}{\partial x}(U_1 + U_2) + 2V_2\right] \tag{2.19}$$

Furthermore, if the two surfaces are rigid, the second term of the right-hand side denoting expansion and contraction of the surfaces becomes zero, and the above equation becomes:

$$\frac{\partial}{\partial x}\left(h^3\frac{\partial p}{\partial x}\right) + \frac{\partial}{\partial z}\left(h^3\frac{\partial p}{\partial z}\right) = 6\mu\left[(U_1 - U_2)\frac{\partial h}{\partial x} + 2V_2\right] \tag{2.20}$$

This is a pressure equation derived based on Reynolds' assumptions and is called **Reynolds' equation**. The previous equations, Eqs. 2.19 and 2.18, can be called generalized Reynolds' equations because they were derived without the rigid body assumption and the isoviscous assumption (in the x and z directions), respectively.

If the flow is one dimensional and $U_2 = V_2 = 0$, the above equation becomes:

$$\frac{\partial}{\partial x}\left(h^3\frac{\partial p}{\partial x}\right) = 6\mu U_1\frac{\partial h}{\partial x} \tag{2.21}$$

This is Reynolds' equation in the simplest case.

2.2.1 Interpretation of Reynolds' Equation

a. Mechanisms of Pressure Generation

Equation 2.19 is shown below for $U_2 = 0$. Let us consider the meaning of Reynolds' equation in this case.

$$\frac{\partial}{\partial x}\left(h^3 \frac{\partial p}{\partial x}\right) + \frac{\partial}{\partial z}\left(h^3 \frac{\partial p}{\partial z}\right) = 6\mu\left[U_1 \frac{\partial h}{\partial x} + h\frac{\partial U_1}{\partial x} + 2V_2\right] \quad (2.22)$$

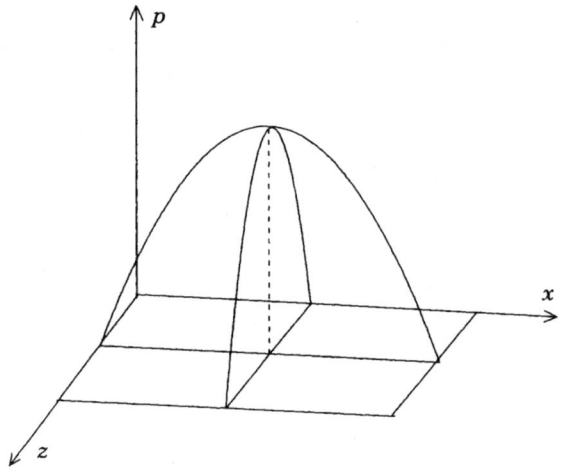

Fig. 2.7. Pressure distribution

First, the left-hand side indicates approximately the average curvature of the pressure distribution surface as shown in Fig. 2.7. If the left-hand side is negative, it means that the pressure distribution is upward convex, or the pressure generated is positive.

Second, the right-hand side represents the causes of pressure generation and the three terms correspond to the following three mechanisms of pressure generation, respectively:

1. The first term represents the **wedge effect**: pressure generation due to the fluid being driven from the thick end to the thin end of the wedge-shaped fluid film by the surface movement.
2. The second term is the **stretch effect**: pressure generation due to the variation of surface velocity from place to place.
3. The third term is the **squeeze effect**: pressure generation due to the variation of surface gap (film thickness).

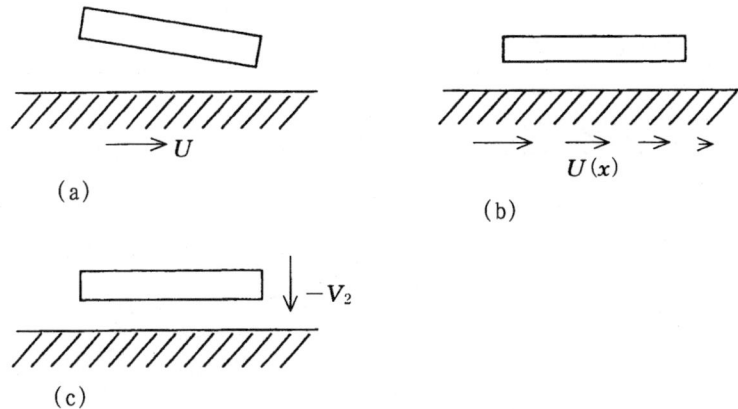

Fig. 2.8a-c. Mechanism of pressure generation. **a** wedge effect, **b** stretch effect, **c** squeeze effect

The simplest examples of these three effects are shown in Fig. 2.8. Each example shows the case where the term is negative, i.e., the pressure is positive.

Among these three, the wedge effect is most commonly seen, for example, in journal bearings. The squeeze effect plays an important role in the small-end bearings of crank rods and is fundamentally important also in animal joints, as pointed out by Reynolds in his paper of 1886 [3]. The stretch effect can be a problem when sliding surfaces are made of elastic materials such as rubber, but can usually be ignored. When a journal moves around in a bearing, an equivalent stretch effect can occur, but it is of the order of the effect of V_2 in Eq. 2.22 multiplied by h/R, and can usually be disregarded. In the case of lubrication during plastic working, for example lubrication between a roller and the workpiece in rolling or lubrication between the die and the workpiece in drawing, deformation of the worked surface becomes a problem. In these cases, since the surface is usually stretched, the generated pressure is negative and so the formation of the lubricant film is hard to explain hydrodynamically.

Now, let us consider Eq. 2.20. It is written again here for the sake of convenience:

$$\frac{\partial}{\partial x}\left(h^3 \frac{\partial p}{\partial x}\right) + \frac{\partial}{\partial z}\left(h^3 \frac{\partial p}{\partial z}\right) = 6\mu \left[(U_1 - U_2)\frac{\partial h}{\partial x} + 2V_2\right] \qquad (2.23)$$

It is interesting to compare the meaning of the right-hand side of this equation in the case of an inclined pad bearing and in the case of a journal bearing.

Consider the following conditions in an inclined pad bearing ($V_2 = 0$) as shown in Fig. 2.9a. RHS stands for "right-hand side."

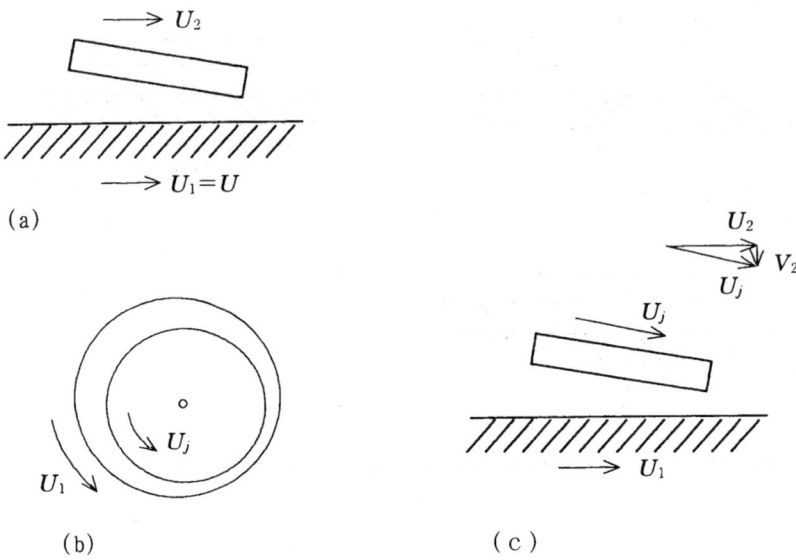

Fig. 2.9a-c. Inclined pad and journal bearings (1). **a** inclined pad bearing, **b** journal bearing, **c** simplified journal bearing

1. If $U_1 = U$ and $U_2 = U$, RHS $= 6\mu(0 + 0) = 0$

2. If $U_1 = U$ and $U_2 = 0$, RHS $= 6\mu\left(U\dfrac{\partial h}{\partial x} + 0\right) = 6\mu U\dfrac{\partial h}{\partial x}$

3. If $U_1 = U$ and $U_2 = -U$, RHS $= 6\mu\left(2U\dfrac{\partial h}{\partial x} + 0\right) = 12\mu U\dfrac{\partial h}{\partial x}$

In case 1, the relative velocity of the two surfaces is 0 and so no pressure is generated. If cases 2 and 3 are compared, the relative velocity in case 3 is twice that in case 2, and accordingly the generated pressure is twice that of case 2.

Turning to the journal bearing shown in Fig. 2.9b, let us expand the cylindrical surface of the journal to a plane as shown in Fig. 2.9c. This is different from Fig. 2.9a in the important point that the circumferential velocity U_j of the journal is along the slope, and this is equivalent to a surface having the velocity of $U_2 \approx U_j$ in the x direction and that of $V_2 \approx U_j(\partial h/\partial x)$ in the y direction. Therefore, the following relations are obtained:

1. If $U_1 = U$ and $U_j = U$, RHS $= 6\mu\left(0 + 2U\dfrac{\partial h}{\partial x}\right) = 12\mu U\dfrac{\partial h}{\partial x}$

2. If $U_1 = 0$ and $U_j = U$, RHS $= 6\mu\left(-U\dfrac{\partial h}{\partial x} + 2U\dfrac{\partial h}{\partial x}\right) = 6\mu U\dfrac{\partial h}{\partial x}$

3. If $U_1 = -U$ and $U_j = U$, RHS $= 6\mu\left(-2U\dfrac{\partial h}{\partial x} + 2U\dfrac{\partial h}{\partial x}\right) = 0$

It is interesting to compare these results with those for the case of Fig. 2.9a. For example, if they are compared for case 2, the result, $6U(\partial h/\partial x)$, is the same, but the factors contributing to this result are different. For the inclined pad bearing, $6U(\partial h/\partial x)$ is directly obtained from the wedge effect. For the journal bearing, $6U(\partial h/\partial x)$ is interpreted as the sum of the wedge effect, which is negative, and the squeeze effect (the effect of V_2), which is positive with a magnitude twice that of the wedge effect.

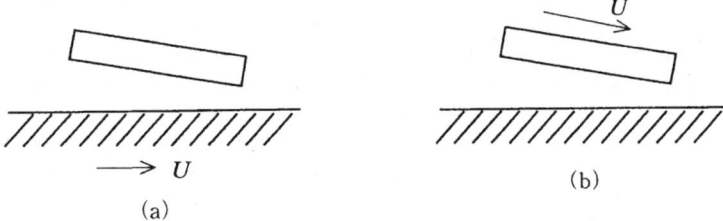

Fig. 2.10a,b. Inclined pad and journal bearings (2). **a** Fig. 2.9a with $U_1 = U$ and $U_2 = 0$, **b** Fig. 2.9c with $U_1 = 0$ and $U_j = U$

Although it can be said from this analysis that the mechanisms of pressure generation in an inclined pad bearing and a journal bearing are different, this is not necessarily true. Figure 2.9a and 2.9c can be redrawn as Fig. 2.10a,ba and 2.10a,bb; considering case 2, it is clear that Fig. 2.10a,bb is equivalent to 2.10a,ba if viewed upside down. Therefore, the difference described above is only an apparent one.

In the case of journal bearings, the circumferential speed of the journal U_2 is along the inclined surface, and Eq. 2.20 becomes:

$$\dfrac{\partial}{\partial x}\left(h^3\dfrac{\partial p}{\partial x}\right) + \dfrac{\partial}{\partial z}\left(h^3\dfrac{\partial p}{\partial z}\right) = 6\mu\left[(U_1 + U_2)\dfrac{\partial h}{\partial x} + 2V_2\right] \tag{2.24}$$

It is this form of the equation that appears in Reynolds' original paper [3].

b. Stationary Surfaces and Moving Surfaces

The upper surface of Fig. 2.10a,ba and the lower surface of Fig. 2.10a,bb are equivalent as stated above, and are called the **stationary surface**. Similarly, the lower surface of 2.10a,ba and the upper surface of 2.10a,bb are equivalent, and they are called the **moving surface**. When 2.10a,ba is regarded as an inclined pad bearing, the upper inclined surface is usually stationary and the lower surface is moving, in agreement with the above nomenclature. However, the meanings of "stationary" and "moving" here are more fundamental, i.e., the stationary surface is a surface in which the distance from a fixed point (on the stationary surface) to the mating surface does not change, and the moving surface is a surface in which the distance from a fixed point (on the moving surface) to the mating surface does change.

The distinction of a stationary surface and a moving surface is important when the temperatures of the two surfaces are different or when the roughness values of the two surfaces are different.

References

1. Beauchamp Tower, "First Report on Friction-Experiments (Friction of Lubricated Bearings)",*Proc. Institution of Mechanical Engineers*, November 1883, pp. 632 - 659.("Adjourned Discussion", January 1884, pp. 29 - 35)
2. Beauchamp Tower, "Second Report on Friction Experiments (Experiments on the Oil Pressure in a Bearing)", *Proc.Institution of Mechanical Engineers*, January 1885, pp. 58 - 70.
3. Osborne Reynolds, "On the Theory of Lubrication and its Application to Mr.B. Tower's Experiments", *Philosophical Transaction of Royal Society of London*, Vol. 177, Pt. 1, 1886, pp. 157 - 234.
4. Oscar Pinkus and Beno Sternlicht, "Theory of Hydrodynamic Lubrication", *McGraw-Hill Book Company*, New York, 1961.
5. Norimune Soda,"Bearings"(in Japanese), *Iwanami Zensho Series, Iwanami Shoten*, Tokyo, 1964.
6. Allastair Cameron, "Principles of Lubrication", *Longmans*, London, 1966.
7. J. Frêne, D. Nicolas, B. Degueurce, D. Berthe, M. Godet, "Lubrification hydrodynamique - Palliers et Butées", *Eyrolles*, Paris, 1990.

3

Fundamentals of Journal Bearings

Oil film bearings are classified by the principle of operation, the direction of loading, and type as follows.

Classification by Operating Principle

– *Hydrodynamic bearings* produce oil film pressure to support a load by shaft rotation inside the bearing.

– *Hydrostatic bearings* also use the oil film pressure to support a load, but the pressure is supplied by an outside source.

Classification by Direction of Loading

– *Journal bearings* support a load acting in the direction normal to the rotating shaft. It is called a journal bearing because the neck of the shaft (the part of the shaft inside the bearing) is called a journal. It is the most common bearing among sliding bearings.

– *Thrust bearings* support a thrust, or a load acting along the axis of the shaft.

Classification of Journal Bearings by Type (see Fig. 3.1a-f)

– For *circular bearings*, the right-angle cross section of the inner surface of bearing metal is a continuous circle. This is the most fundamental form of journal bearing.

– *Partial bearings* have semicircular bearing metal only on the loaded side of the journal. They are superior to a full circular bearing because of reduced frictional losses. They are frequently used to support large static loads, e.g., large turbines. A stopper is placed on the other side of the journal for safety.

– *Two arc bearings*, and *three arc bearings* have a cross section that is a combination of two or three circular arcs (lobes). The general term for these types of bearings is multi-arc bearings. These bearings are used for better stability of rotating shafts (see Chapter 5).

– *Tilting pad bearings* consist of several pivoted pads that can tilt freely. The stability of rotating shafts can be improved by using this type of bearing. Oil whip can be suppressed almost completely. This is especially useful for light load, high speed rotating shafts, which tend to be unstable (see Chapter 5).

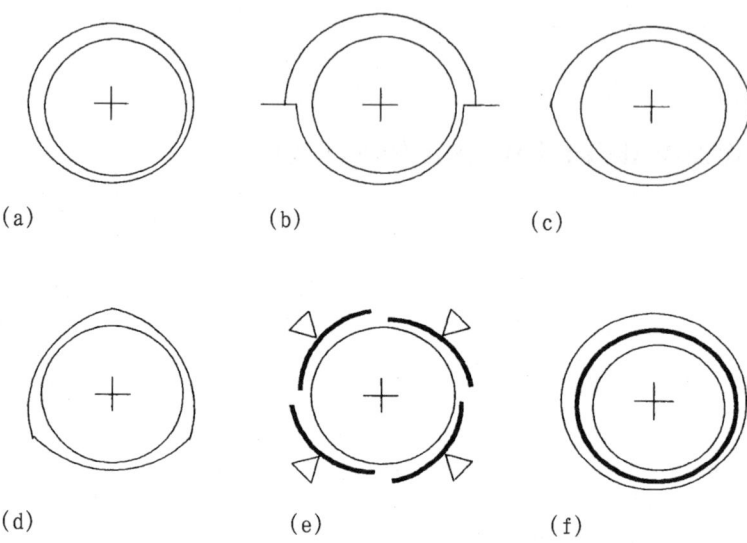

Fig. 3.1a-f. Forms of journal bearing. **a** circular bearing, **b** partial bearing, **c** two arc bearing, **d** three arc bearing, **e** tilting pad bearing, **f** floating bush bearing

– *Floating bush bearings* have a freely rotating floating bush (thin metal cylinder) between the journal and the bearing metal. Such bearings are used to suppress shaft vibrations (see Chapter 5).

Classification by Lubricant
– Bearings are classified by lubricant type into oil bearings, gas bearings, and grease bearings among others.

Journal bearings are used even today as indispensable bearings in many rotating machines such as steam turbines, generators, blowers, compressors, internal combustion engines, and ship propulsion shafts, even though rolling element bearings are widely used. This is because journal bearings are superior to rolling element bearings in vibration absorption, shock resistance, quietness, and long life. All these characteristics come from the journal bearing principle of supporting a shaft by a thin oil film. Also, the smaller outside diameter of a journal bearing compared with a rolling element bearing is often beneficial to designers.

This chapter is an introduction to the static characteristics of hydrodynamic circular journal bearings using oil as a lubricant [5].

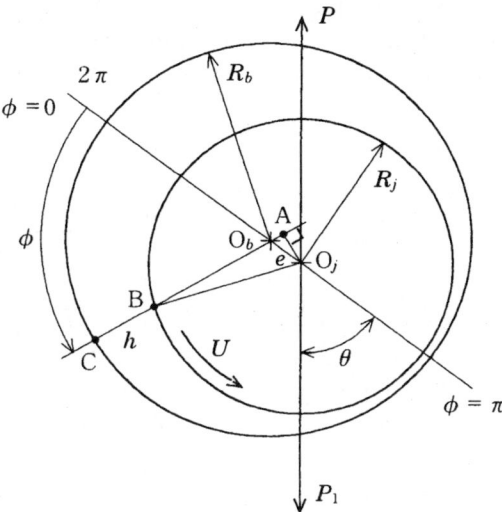

Fig. 3.2. Full circular journal bearing

3.1 Circular Journal Bearings

3.1.1 Cross Section of a Bearing

Figure 3.2 shows the cross section of a full circular journal bearing. O_b is the center of the bearing, O_j is the center of the journal, R_b is the inner radius of the bearing metal (bush), R_j is the radius of the journal, and U is the circumferential velocity of the journal surface. The difference between the inner radius of the bearing and the radius of the journal, $c = R_b - R_j$, is called the **radial clearance** of the bearing. Although the radial clearance is very small compared with the bearing radius (usually less than 1/500), the clearance is very much exaggerated in the figure. Oil, which fills the bearing clearance, is called the **oil film**. P is the resultant of the oil pressure in the oil film and P_1 is the load acting on the journal. P and P_1 are in equilibrium.

The center of the journal rotating in a bearing under a load is shifted from the bearing center in a diagonally downward direction, making an angle θ with the direction of loading (taken downward). The angle θ measured from the loading direction to the shifting direction is called the **attitude angle**. The distance from the bearing center to the journal center, $e = \overline{O_b O_j}$, is called the **eccentricity**, and its ratio to the bearing clearance, $\kappa = e/c$, is called the **eccentricity ratio**. $\kappa = 0$ means that the journal and the bearing are concentric and $\kappa = 1$ means that the journal is in contact with the bearing metal.

The motion of the journal center is confined within a circle with a radius equal to the radial clearance c with its center at the bearing center. Or, it can be stated as being confined within a circle of radius $\kappa = 1$, if the eccentricity ratio is used. This circle, shown in Fig. 3.3a,ba, is called the **clearance circle**. The clearance circle is conveniently used to show the relative position of the journal center to the bearing center. When the motion of the journal is confined to a certain quadrant, it is sufficient to draw only the necessary quadrant of the clearance circle, as shown in Fig. 3.3a,bb.

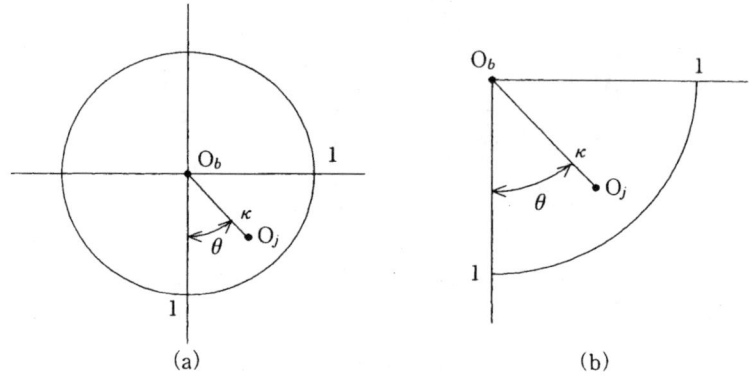

Fig. 3.3a,b. Clearance circle. **a** whole circle, **b** the relevant quadrant

3.1.2 Shape of the Oil Film

In the range $\phi = 0 - \pi$ in Fig. 3.2, the oil film thickness decreases in the direction of journal rotation, whereas it increases in the range $\phi = \pi - 2\pi$. In the range where the film thickness decreases in the direction of journal rotation, positive pressure occurs in the oil film due to the journal rotation (wedge effect), and this supports the load on the journal. Although negative pressure conversely occurs in the region where the oil film thickness increases, its absolute value is usually small. Pressure generation greatly depends on the shape of oil film (distribution of oil film thickness).

Let us first obtain the **oil film thickness** (bearing clearance) h below. In Fig. 3.2, the straight line ABC is drawn through O_b. Let A denote the foot of the perpendicular dropped from O_j to the above straight line, and B and C denote the intersections of the straight line and the journal surface and the inner surface of bearing metal, respectively. Then:

$$h = \overline{BC} = \overline{AC} - \overline{AB}$$

where

$$\overline{AC} = R_b + e \cos \phi, \quad \overline{AB} = \sqrt{R_j^2 - (e \sin \phi)^2}$$

Therefore,

$$h = R_b + e \cos \phi - R_j \sqrt{1 - (e/R_j)^2 \sin^2 \phi}$$

Since e/R_j is usually a small value of the order of 1/500 or less, disregarding the second term in the square root, we have:

$$h = (R_b - R_j) + e\cos\phi = c + e\cos\phi$$

Thus, oil film thickness h is obtained with good approximation as:

$$h = c(1 + \kappa \cos\phi) \tag{3.1}$$

3.1.3 Bearing Length (Bearing Width)

The oil film pressure in a journal bearing in the stationary state is given by the following Reynolds' equation, which is derived from Eq. 2.20 or Eq. 2.24, where the x axis is taken in the circumferential direction of the bearing and the z axis is in the axial direction:

$$\frac{\partial}{\partial x}\left(h^3 \frac{\partial p}{\partial x}\right) + \frac{\partial}{\partial z}\left(h^3 \frac{\partial p}{\partial z}\right) = 6\mu U_2 \frac{\partial h}{\partial x} \tag{3.2}$$

where $p(x,z)$ is the oil film pressure, $h(x,z)$ is the oil film thickness, and U_2 is the circumferential velocity of the journal.

General solutions of Eq. 3.2 cannot be obtained analytically, therefore, the following approximations or numerical solutions are usually employed.

– The *infinite length approximation* assumes a sufficiently long bearing in the axial direction and neglects the second term of the left-hand side of the above equation, allowing it to be solved analytically.

– The *short bearing approximation* assumes a sufficiently short bearing in the axial direction and neglects the first term of the left-hand side of the above equation, allowing it to be solved analytically.

– For *finite length bearings*, the equation is solved numerically by, for example, the finite difference method or the finite element method, or by an approximate analytical method by developing the pressure as a series of trigonometric functions.

3.1.4 Boundary Conditions for the Oil Film

Boundary conditions are required to solve Reynolds' equation, Eq. 3.2. In the case of a journal bearing, the boundary condition at an end of the bearing is simply that the oil film pressure is equal to ambient air pressure, because the boundary of the oil film at the end of the bearing is clear-cut. In the circumferential direction, however, since rupture of the oil film in the domain where the clearance is growing is a complicated phenomenon (for various reasons including the inflow of air), it is a difficult problem to determine the boundary of the oil film.

For simplicity, consider an infinitely long journal bearing. In this case, if it is assumed that the bearing clearance is completely filled with oil (no film rupture assumed), Reynolds' equation gives positive pressure in the semicircle where the bearing clearance decreases and negative pressure in the semicircle where the bearing clearance increases, and their absolute values are equal. This will be true if the bearing pressure is sufficiently low. When the bearing pressure is relatively high, however, although the positive pressure can go up without limitation, the negative pressure cannot go below a certain limit. When the absolute value of negative pressure reaches some limit, oil film rupture occurs and the pressure in the region of rupture will not go down further. It is difficult to know exactly the position at which oil film rupture will occur (the oil film terminal point) and the pressure in the area of oil film rupture.

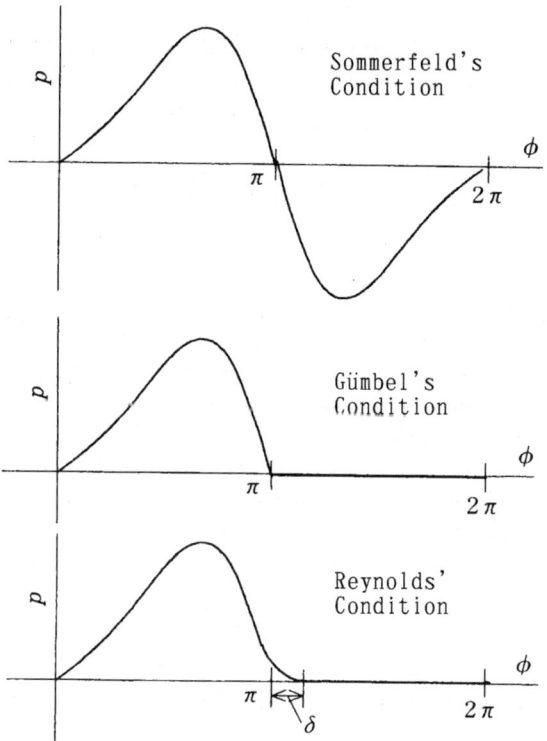

Fig. 3.4. Boundary conditions for the oil film

Simplified boundary conditions, as shown in Fig. 3.4, are often used in practice.

(a) *Sommerfeld's boundary condition* (A. J. W. Sommerfeld, 1869–1951) assumes that $p = 0$ at $\phi = 0$ and 2π. Pressure distribution is calculated without considering oil film rupture, and the positive and negative pressures obtained are taken into consideration as they are. This is applicable when the bearing pressure is low.

(b) For *Gümbel's boundary condition* (L. K. F. Gümbel, 1874–1923), pressure distribution is calculated without considering oil film rupture as before, but only the positive pressure in the semicircle $\phi = 0 - \pi$ is considered. The negative pressure in the remaining semicircle is set to "zero" (i.e., atmospheric pressure). The oil film is thus assumed to terminate at $\phi = \pi$. The starting position is $\phi = 0$. This is acceptable when the bearing pressure is fairly high. It is also called the half Sommerfeld's condition.

(c) In *Reynolds' boundary condition* (O. Reynolds, 1842–1912), the oil film is assumed to terminate at a certain position ($\phi = \pi + \delta$) at which both the pressure and pressure gradient are zero, simultaneously. This condition eliminates a discontinuity of oil flow at $\phi = \pi$, a physical contradiction involved in Gümbel's condition. It is now necessary, however, to determine δ. This is also known as Swift–Stieber's condition.

(d) In *the boundary condition of separation*, the terminal position of the oil film is determined from the separation conditions of flow [7] [8].

(e) Another approach divides the cause of pressure generation into the wedge effect and the squeeze effect and applies the boundary condition $p = 0$ at $\phi = (0, \pi)$ and $p = 0$ at $\phi = (\pi/2, 3\pi/2)$ to the two effects, separately.

3.2 Infinitely Long Bearings

The basic characteristics of an infinitely long journal bearing are derived here. Reynolds' equation is solved first to determine the oil film pressure, which is then integrated to obtain the oil film force. From the equilibrium of the oil film force and bearing load, values for the eccentricity ratio, load capacity, and frictional moment are calculated.

3.2.1 Oil Film Pressure

In the case of an infinitely long bearing, it can be assumed that the oil film pressure does not change in the axial direction, or $(\partial p/\partial z) = 0$. Therefore, with $U = U_2$, Eq. 3.2 can be simplified as follows:

$$\frac{d}{dx}\left(h^3 \frac{dp}{dx}\right) = 6\mu U \frac{dh}{dx} \tag{3.3}$$

Integrating this with respect to x yields:

$$h^3 \frac{dp}{dx} = 6\mu U (h + C_1) \tag{3.4}$$

where C_1 is an integral constant. Letting the film thickness at the maximum film pressure position (where $dp/dx = 0$) be h_m, we have $C_1 = -h_m$. Note that h_m is still unknown. Using h_m in the above equation leads to:

$$\frac{dp}{dx} = 6\mu U \left(\frac{1}{h^2} - \frac{h_m}{h^3}\right) \tag{3.5}$$

Substituting Eq. 3.1 into the above equation gives

$$\frac{dp}{d\phi} = \frac{6\mu U R}{c^2}\left[\frac{1}{(1+\kappa \cos \phi)^2} - \frac{h_m}{c}\frac{1}{(1+\kappa \cos \phi)^3}\right] \tag{3.6}$$

where, since $R_b \approx R_j$, both the radii are denoted by R and the relation $x = R\phi$ is used.

To obtain the oil film pressure, Eq. 3.6 is integrated again to give the following, in which C_2 is an integral constant:

$$p = \frac{6\mu UR}{c^2} \left[\int \frac{d\phi}{(1 + \kappa \cos \phi)^2} - \frac{h_m}{c} \int \frac{d\phi}{(1 + \kappa \cos \phi)^3} \right] + C_2 \quad (3.7)$$

To calculate the integrals on the right-hand side of the above equation, Sommerfeld introduced the following variable transform [1][5] (another method is given in Section 5.2.1):

$$1 + \kappa \cos \phi = \frac{1 - \kappa^2}{1 - \kappa \cos \varphi} \quad (3.8)$$

where φ is the new variable. In this variable transform, the range $\phi = 0 - 2\pi$ corresponds to the same range $\varphi = 0 - 2\pi$, and this makes it easy to handle the boundary conditions.

First, $\cos \phi$ is obtained from Eq. 3.8, and then $\sin \phi$ is found from the relations $\sin^2 \phi + \cos^2 \phi = 1$, as follows:

$$\cos \phi = \frac{\cos \varphi - \kappa}{1 - \kappa \cos \varphi}, \quad \sin \phi = \frac{(1 - \kappa^2)^{1/2} \sin \varphi}{1 - \kappa \cos \varphi} \quad (3.9)$$

From these relations, we have the following:

$$d\phi = \frac{(1 - \kappa^2)^{1/2}}{1 - \kappa \cos \varphi} d\varphi \quad (3.10)$$

With these relations, the integrals in Eq. 3.7 are calculated as follows (J_1 will be used later):

$$J_3 = \int \frac{d\phi}{(1 + \kappa \cos \phi)^3} = \frac{1}{(1 - \kappa^2)^{5/2}} \left(\varphi - 2\kappa \sin \varphi + \frac{\kappa^2 \varphi}{2} + \frac{\kappa^2}{4} \sin 2\varphi \right) \quad (3.11)$$

$$J_2 = \int \frac{d\phi}{(1 + \kappa \cos \phi)^2} = \frac{1}{(1 - \kappa^2)^{3/2}} \left(\varphi - \kappa \sin \varphi \right) \quad (3.12)$$

$$J_1 = \int \frac{d\phi}{1 + \kappa \cos \phi} = \frac{\varphi}{(1 - \kappa^2)^{1/2}} \quad (3.13)$$

Now, Eq. 3.7, the pressure distribution, becomes:

$$p(\varphi) = \frac{6\mu UR}{c^2} \left[\frac{1}{(1 - \kappa^2)^{3/2}} \left(\varphi - \kappa \sin \varphi \right) \right.$$
$$\left. - \frac{h_m}{c(1 - \kappa^2)^{5/2}} \left(\varphi - 2\kappa \sin \varphi + \frac{\kappa^2 \varphi}{2} + \frac{\kappa^2}{4} \sin 2\varphi \right) \right] + C_2 \quad (3.14)$$

where h_m and C_2 are integral constants, which can be determined under appropriate boundary conditions. Calculations will be continued hereafter under typical boundary conditions.

3.2.2 Infinitely Long Bearing Under Sommerfeld's Condition

a. Oil Film Pressure

The basic characteristics of an infinitely long journal bearing are investigated under Sommerfeld's boundary condition. The boundary conditions used here are as follows:

$$p = 0 \text{ at } \phi = 0 \text{ and } \phi = 2\pi \tag{3.15}$$

where the ambient pressure is assumed to be $p = 0$. As mentioned above, the boundary conditions with respect to φ are also:

$$p = 0 \text{ at } \varphi = 0 \text{ and } \varphi = 2\pi \tag{3.16}$$

where the integral constants C_2 and h_m are determined as follows:

$$C_2 = 0 \tag{3.17}$$

$$h_m = \frac{2(1 - \kappa^2)}{2 + \kappa^2} c \tag{3.18}$$

Substituting these values into Eq. 3.14 and returning the variable φ to ϕ gives the following pressure distribution.

$$p(\phi) = \frac{6\mu U R}{c^2} \frac{\kappa(2 + \kappa \cos \phi) \sin \phi}{(2 + \kappa^2)(1 + \kappa \cos \phi)^2} \tag{3.19}$$

$$\equiv \frac{6\mu U R}{c^2} \bar{p}(\kappa, \phi) \tag{3.20}$$

The nondimensional pressure distribution function $\bar{p}$ in Eq. 3.20 is shown against ϕ in Fig. 3.5. The parameter is the eccentricity ratio κ.

As shown in Fig. 3.5, the pressure distribution is symmetric with respect to the point ($\phi = \pi$, $\bar{p} = 0$) and the absolute values of the highest and the lowest pressures are equal. The position ϕ_0 of these extrema are determined from the condition $dp/d\phi = 0$ as follows.

$$\cos \phi_0 = \frac{-3\kappa}{2 + \kappa^2} \tag{3.21}$$

This shows that, with increase in the eccentricity ratio κ from 0 to 1, the position of the highest pressure ϕ_0 moves in the direction of rotation of the journal from $\pi/2$ toward π, while that of the minimum pressure ϕ_0 moves in the opposite direction from $3\pi/2$ toward π.

b. Oil Film Force and Load Capacity

Integration of the oil film pressure gives the oil film force P. This must balance the bearing load P_1. The balance of the forces, if resolved into two directions (one in the

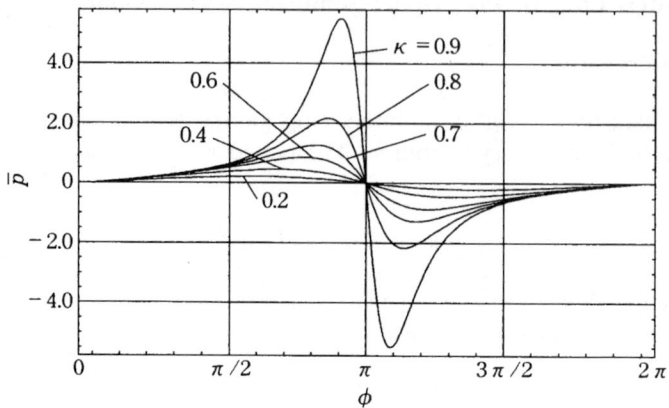

Fig. 3.5. Pressure distribution in an infinitely long bearing using Sommerfeld's condition

eccentricity direction and the other perpendicular to it), can be expressed as follows (see Fig. 3.6):

$$LR \int_0^{2\pi} p \cos \phi \, d\phi + P_1 \cos \theta = 0 \tag{3.22}$$

$$LR \int_0^{2\pi} p \sin \phi \, d\phi - P_1 \sin \theta = 0 \tag{3.23}$$

where L and R are the length and the radius of the bearing, respectively, and p is the oil film pressure given by Eq. 3.19.

If Eq. 3.22 is integrated by parts, we have:

$$P_1 \cos \theta = -LR \left[p \sin \phi \right]_0^{2\pi} + LR \int_0^{2\pi} \frac{dp}{d\phi} \sin \phi \, d\phi$$

Since the pressure p is finite, the first term of the right-hand side is clearly zero. The second term can be calculated by using Eq. 3.6, and will turn out to be zero as follows:

$$\text{2nd term} = 6\mu UL \left(\frac{R}{c}\right)^2 \left[\int_0^{2\pi} \frac{\sin \phi}{(1 + \kappa \cos \phi)^2} d\phi - \frac{h_m}{c} \int_0^{2\pi} \frac{\sin \phi}{(1 + \kappa \cos \phi)^3} d\phi \right]$$

$$= 6\mu UL \left(\frac{R}{c}\right)^2 \left[\frac{1}{\kappa(1 + \kappa \cos \phi)} - \frac{h_m}{c} \frac{1}{2\kappa(1 + \kappa \cos \phi)^2} \right]_0^{2\pi} = 0$$

Therefore,

$$P_1 \cos \theta = 0 \tag{3.24}$$

If $P_1 \neq 0$, the attitude angle θ will be determined as follows in the range of $(0 - \pi)$.

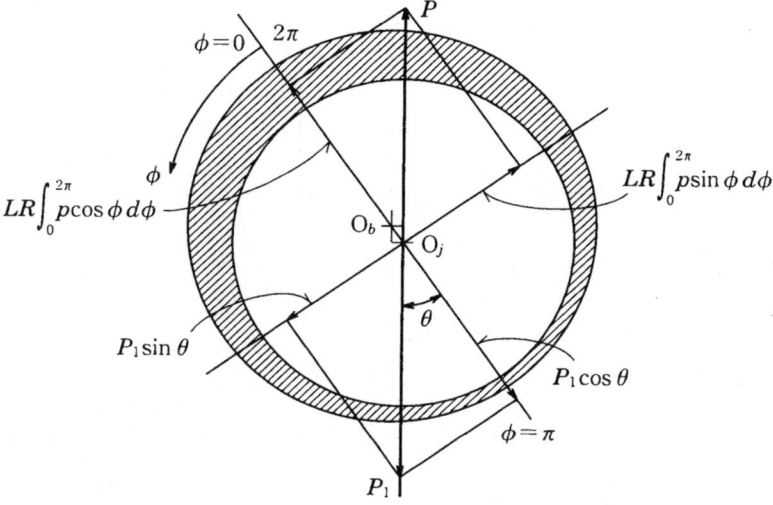

Fig. 3.6. Oil film and oil film force under Sommerfeld's condition

$$\theta = \frac{\pi}{2} \tag{3.25}$$

This means that the bearing load P_1 and the line of the eccentricity are mutually perpendicular, as shown in Fig. 3.7. In this case, the locus of the journal center is a straight line that is normal to the bearing load P_1.

The fact that the oil film force has a component perpendicular to the eccentricity direction (in this case only a perpendicular component) is one of the most unfavorable characteristics of the oil film force in a journal bearing and it causes whirling of the shaft (see Chapter 5).

If Eq. 3.23 is similarly integrated by parts, since $P_1 \sin\theta = P_1$, we have:

$$P_1 = -LR\Big[p\cos\phi\Big]_0^{2\pi} + LR\int_0^{2\pi} \frac{dp}{d\phi}\cos\phi\, d\phi$$

$$= 6\mu UL\left(\frac{R}{c}\right)^2\left[\int_0^{2\pi}\frac{\cos\phi}{(1+\kappa\cos\phi)^2}d\phi - \frac{h_m}{c}\int_0^{2\pi}\frac{\cos\phi}{(1+\kappa\cos\phi)^3}d\phi\right]$$

The integrands on the right-hand side are expanded into partial fractions as follows:

$$\frac{\cos\phi}{(1+\kappa\cos\phi)^2} = \frac{1}{\kappa}\left[\frac{1}{(1+\kappa\cos\phi)} - \frac{1}{(1+\kappa\cos\phi)^2}\right]$$

$$\frac{\cos\phi}{(1+\kappa\cos\phi)^3} = \frac{1}{\kappa}\left[\frac{1}{(1+\kappa\cos\phi)^2} - \frac{1}{(1+\kappa\cos\phi)^3}\right]$$

Integrating these functions with recourse to Sommerfeld's transform of variables gives P_1:

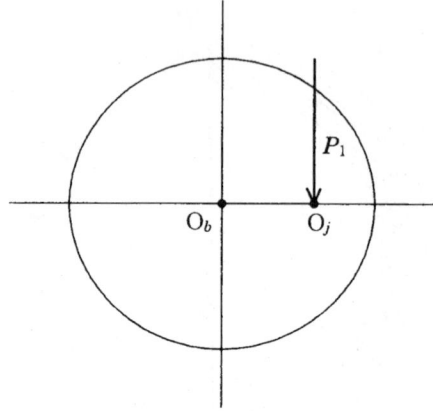

Fig. 3.7. Position of the journal center in an infinitely long bearing under Sommerfeld's condition

$$P_1 = 6\mu UL \left(\frac{R}{c}\right)^2 \frac{1}{\kappa} \left[\frac{h_m}{c} J_3 - \left(\frac{h_m}{c} + 1\right) J_2 + J_1 \right]_0^{2\pi} \tag{3.26}$$

Substituting J_3, J_2, J_1, and h_m previously obtained into the above equation gives the bearing load P_1 as follows:

$$P_1 = P_{1\theta} = \mu UL \left(\frac{R}{c}\right)^2 \frac{12\pi\kappa}{(2+\kappa^2)(1-\kappa^2)^{1/2}} \tag{3.27}$$

This equation can be rewritten using the average bearing pressure $p_m = P_1/(2RL)$ and the number of revolutions of the shaft per unit time $N = U/(2\pi R)$:

$$\frac{p_m}{\mu N} \left(\frac{c}{R}\right)^2 = \frac{12\pi^2 \kappa}{(2+\kappa^2)(1-\kappa^2)^{1/2}} \tag{3.28}$$

The left-hand side of Eq. 3.28 is a combination of the dimensions of the bearing and some variables representing its operating conditions, and is nondimensional as a whole. The right-hand side is a nondimensional function of the eccentricity ratio κ only. The reciprocal of the nondimensional quantity on the left-hand side is called **Sommerfeld's number** and is usually denoted by S, i.e.,

$$S \equiv \frac{\mu N}{p_m} \left(\frac{R}{c}\right)^2 \tag{3.29}$$

Equation 3.28 can be rewritten as follows using S:

$$\frac{1}{S} = \frac{12\pi^2 \kappa}{(2+\kappa^2)(1-\kappa^2)^{1/2}} \tag{3.30}$$

The relation between $1/S$ and κ is shown in Fig. 3.8. If the bearing dimensions, oil viscosity, bearing pressure, and the number of journal revolutions per unit time are given, Sommerfeld's number S is determined by Eq. 3.29, and the corresponding eccentricity ratio κ can be determined by Eq. 3.30 or from Fig. 3.8. Since the same Sommerfeld's number gives the same eccentricity ratio κ even though the bearing dimensions and operating conditions are different, it can be said that Sommerfeld's number is an important quantity relating to the similarity rule for a bearing. Further, the definition of Sommerfeld's number, Eq. 3.29, shows how the various factors of a bearing are related to the operating condition of a bearing.

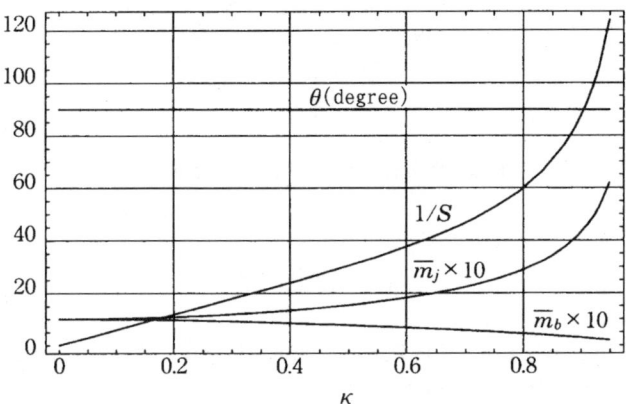

Fig. 3.8. Bearing characteristics for an infinitely long bearing under Sommerfeld's conditions. S, Sommerfeld's number; $\bar{m}_j$ and $\bar{m}_b$, nondimensinal, frictional moments; θ, attitude angle

c. Minimum Clearance and Load Capacity

If the eccentricity ratio κ is thus determined, the minimum clearance of the bearing is obtained from Eq. 3.1 as:

$$h_{\min} = c\,(1 - \kappa) \tag{3.31}$$

Then, the bearing load P_1 corresponding to the allowable minimum value of $h_{\min}$ (this depends on the surface roughness and machining accuracy among other factors) will be the maximum allowable load (load capacity) of the bearing.

d. Frictional Moment and Frictional Loss

The frictional moment due to the shear stress in the oil film acts on the rotating journal. Its reaction acts on the bearing bush.

If the journal and the bearing are concentric (Fig. 3.9), the frictional moment M is simply given as follows, multiplying the shear stress $\tau = \mu(U/c)$ by the radius R and the circumferential area $2\pi RL$.

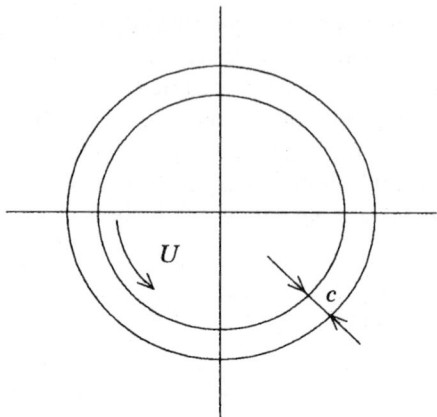

Fig. 3.9. Derivation of Petrov's formula

$$M = \frac{2\pi\mu U R^2 L}{c} \tag{3.32}$$

This is called **Petrov's law** (N.P. Petrov, 1836–1920) and is convenient for a rough estimation of the frictional moment.

If the journal and the bearing are eccentric, the frictional moment on the journal can be calculated by integrating the oil film shear stress over the journal surface and that on the bearing bush by integrating the oil film shear stress over the bush surface. The shear stress is calculated as follows by using Eq. 2.10 for the flow velocity:

$$\tau = \mu\frac{\partial u}{\partial y} = \mu\frac{\partial}{\partial y}\left[\left[\left(1 - \frac{y}{h}\right)U_1 + \frac{y}{h}U_2\right] - \frac{1}{2\mu}\frac{\partial p}{\partial x}y(h-y)\right]$$

$$= -\mu\frac{U_1 - U_2}{h} - \frac{h}{2}\frac{\partial p}{\partial x}\left(1 - \frac{2y}{h}\right)$$

Let $U_1 = 0$ and $U_2 = U$. Then the shear stresses at $y = h$ and $y = 0$ are obtained as follows:

$$\tau_{y=h} = \frac{\mu U}{h} + \frac{h}{2}\frac{dp}{dx} \tag{3.33}$$

$$\tau_{y=0} = \frac{\mu U}{h} - \frac{h}{2}\frac{dp}{dx} \tag{3.34}$$

Multiplying these by the bearing radius and integrating them over the corresponding circumferential surface gives the frictional moment acting on the journal and the bearing bush as follows, where J_1 and J_2 (Eq. 3.13, etc.) are used:

$$M_j = \int_0^{2\pi}\left(\frac{\mu U}{h} + \frac{h}{2R}\frac{dp}{d\phi}\right)R^2 L\,d\phi$$

$$= \frac{2\pi\mu UR^2 L}{c} \frac{2(1+2\kappa^2)}{(2+\kappa^2)(1-\kappa^2)^{1/2}} \equiv M \cdot \bar{m}_j(\kappa) \tag{3.35}$$

$$M_b = \int_0^{2\pi} \left(\frac{\mu U}{h} - \frac{h}{2R}\frac{dp}{d\phi}\right) R^2 L \, d\phi$$

$$= \frac{2\pi\mu UR^2 L}{c} \frac{2(1-\kappa^2)^{1/2}}{2+\kappa^2} \equiv M \cdot \bar{m}_b(\kappa) \tag{3.36}$$

where M is the frictional moment M of Petrov's law, Eq. 3.32, and $\bar{m}_j(\kappa)$ and $\bar{m}_b(\kappa)$ are nondimensional moments which are functions of κ only. If $\kappa = 0$, obviously $M_j = M_b = M$. Equation 3.25, $\bar{m}_j(\kappa)$ and $\bar{m}_b(\kappa)$ are shown in Fig. 3.8.

It should be noted that M_j and M_b are not equal, M_j being always larger than M_b. Calculation shows that the difference is given by:

$$M_j - M_b = eP_1 \tag{3.37}$$

This shows that the difference of M_j and M_b can be attributed to the moment of the bearing load P_1 with respect to the bearing center, the eccentricity being e.

If the frictional moment on the journal is known, the frictional loss (heat generation) L_s is calculated as follows with the angular velocity ω:

$$L_s = \omega M_j \tag{3.38}$$

The frictional coefficients at the journal surface and the inner surface of bearing bush are defined as follows.

$$f_j = M_j/(RP_1), \quad f_b = M_b/(RP_1) \tag{3.39}$$

3.2.3 Infinitely Long Bearing Under Gümbel's Condition

In the previous section, various characteristics of an infinitely long bearing were derived under Sommerfeld's boundary condition. However, one of the results, that the locus of the journal center is a straight line perpendicular to the load direction (see Fig. 3.7), contradicts actual observation in many practical cases. In fact, the locus is a straight line only when the bearing pressure is very low. Under usual conditions, it is rather like the profile of a half-moon. This disagreement seems to be attributable to the inclusion of the large negative pressure from the theory into the calculation of the oil film force. In the actual oil film, the negative pressure cannot be very low because of oil film rupture and air inflow from the bearing ends. In this section, assuming no negative pressure exists in the oil film, let us, for simplicity, adopt Gümbel's boundary condition in which the negative pressure in the theory is replaced by zero, and calculate various characteristics of a bearing under this condition.

a. Oil Film Pressure

Under Gümbel's boundary condition, only the positive pressure in the shaded region of Fig. 3.10, i.e., the range $0 \leq \phi \leq \pi$, is considered. The negative pressure in the range $\pi \leq \phi \leq 2\pi$ is simply assumed to be zero. The pressure in the region of positive pressure is assumed to be the same as that obtained under Sommerfeld's condition.

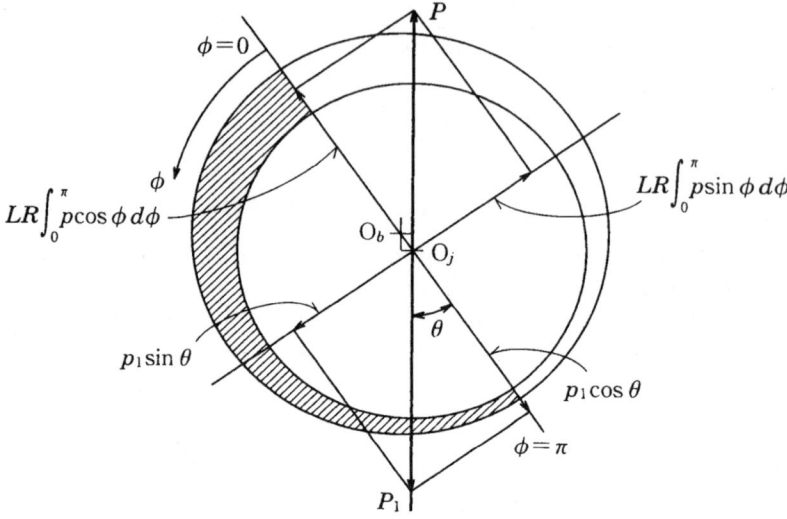

Fig. 3.10. The oil film and oil film force under Gümbel's boundary condition

b. Oil Film Force and Load Capacity

The oil film force under Gümbel's boundary condition is obtained by integrating the oil film pressure p of Eq. 3.19 over the range $0 \leq \phi \leq \pi$. Then, the balance of the oil film force and the bearing load can, when resolved into components in the direction of eccentricity and that perpendicular to it, be written as follows:

$$LR \int_0^\pi p \cos \phi \, d\phi + P_1 \cos \theta = 0 \qquad (3.40)$$

$$LR \int_0^\pi p \sin \phi \, d\phi - P_1 \sin \theta = 0 \qquad (3.41)$$

where L and R are the length and the radius of the bearing metal, respectively. The integration can be performed in the same way as in the previous section, giving the following results:

$$P_1 \cos \theta = \mu U L \left(\frac{R}{c}\right)^2 \frac{12\kappa^2}{(2+\kappa^2)(1-\kappa^2)} \qquad (3.42)$$

$$P_1 \sin \theta = \mu U L \left(\frac{R}{c}\right)^2 \frac{6\pi\kappa}{(2+\kappa^2)(1-\kappa^2)^{1/2}} \qquad (3.43)$$

It is interesting to compare these results with those obtained under Sommerfeld's condition in the previous section. Under Sommerfeld's condition, the positive pressure and the negative pressure generated in the oil film cancel each other out in the direction of eccentricity, giving $P_1 \cos \theta = 0$, whereas in the direction perpendicular to it, they assist each other, giving the large value $P_1 \sin\theta$. Under Gümbel's condition, in contrast, since the negative pressure is assumed to be zero, no cancelling takes place in the direction of eccentricity, leaving $P_1 \cos\theta \neq 0$, and in the direction perpendicular to it, $P_1 \sin\theta$ becomes one-half of that under Sommerfeld's condition.

From Eq. 3.42 and Eq. 3.43, the bearing load P_1 is obtained as:

$$P_1 = \mu U L \left(\frac{R}{c}\right)^2 \frac{6\kappa\{4\kappa^2 + \pi^2(1-\kappa^2)\}^{1/2}}{(2+\kappa^2)(1-\kappa^2)}. \tag{3.44}$$

This can be rewritten as follows by using the Sommerfeld number, S:

$$\frac{1}{S} = \frac{6\pi\kappa\{4\kappa^2 + \pi^2(1-\kappa^2)\}^{1/2}}{(2+\kappa^2)(1-\kappa^2)} \tag{3.45}$$

where $S = (\mu N/p_m)(R/c)^2$ as before, with $N = U/(2\pi R)$ and $p_m = P_1/(2RL)$.

Further, dividing Eq. 3.43 by Eq. 3.42 gives the relation between the attitude angle θ and the eccentricity ratio κ:

$$\tan\theta = \frac{\pi(1-\kappa^2)^{1/2}}{2\kappa} \tag{3.46}$$

This is a polar coordinates expression of the journal center locus as shown in Fig. 3.11. This is like a half-moon and is close to the actual shape of the locus.

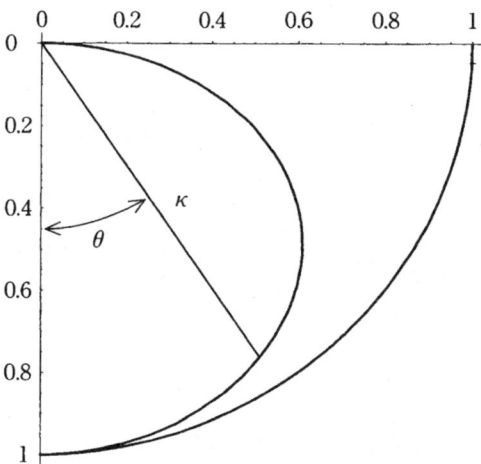

Fig. 3.11. Locus of the journal center for a bearing of infinite length under Gümbel's boundary condition

c. Frictional Moment

Although the oil film exists in the whole circumference of the bearing, if $(\partial p/\partial \phi) = 0$ is assumed in the range $\pi \leq \phi \leq 2\pi$, the frictional moments acting on the journal and the bearing metal are found as follows, in the same way as for Eq. 3.35 and Eq. 3.36:

$$M_j = \frac{2\pi\mu U R^2 L}{c} \frac{1}{2(1-\kappa^2)^{1/2}} \left(2 + \frac{3\kappa^2}{2+\kappa^2}\right) \equiv M\bar{m}_j(\kappa) \qquad (3.47)$$

$$M_b = \frac{2\pi\mu U R^2 L}{c} \frac{1}{2(1-\kappa^2)^{1/2}} \left(2 - \frac{3\kappa^2}{2+\kappa^2}\right) \equiv M\bar{m}_b(\kappa) \qquad (3.48)$$

where M is the M in Petrov's law, Eq. 3.32. In this case also, M_j and M_b are not equal and their difference is found to be equal to the moment of the bearing load with respect to the bearing center, as in the previous section. The relation in this case is:

$$M_j - M_b = (e \sin \theta) P_1 \qquad (3.49)$$

Whereas $\theta = \pi/2$ in the previous section, θ is now a function of κ.

The frictional coefficients at the journal surface and the bearing metal surface are defined as:

$$f_j = M_j/(RP_1), \quad f_b = M_b/(RP_1) \qquad (3.50)$$

When the journal and the bearing metal are concentric, i.e., if $\kappa = 0$, we have:

$$M_j = M_b = M \qquad (3.51)$$

Equations 3.45 – 3.48 are shown in Fig. 3.12.

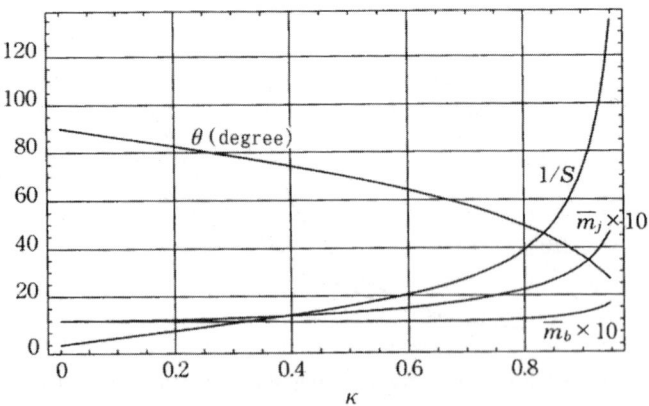

Fig. 3.12. Characteristics of an infinitely long bearing under Gümbel's condition

3.3 Short Bearings

So far, only infinitely long bearings have been considered. In the past, most bearings actually had a length that was more than twice the diameter. In recent years, however, more and more short bearings are being used to reduce uneven axial load distribution and frictional loss, among others. In many cases, the (length (width) / diameter) ratio of a bearing is actually 0.5 – 1. If the bearing length is sufficiently small compared with the diameter, the pressure gradient in the axial direction is much larger than that in the circumferential direction, and the latter can be disregarded. In this case, Reynolds' equation can be written approximately as follows, omitting the first term of the left-hand side of Eq. 3.2:

$$\frac{\partial}{\partial z}\left(h^3 \frac{\partial p}{\partial z}\right) = 6\mu U \frac{dh}{dx} \qquad (3.52)$$

where $U = U_2$. Such an approximation is called the DuBois–Ocvirk short bearing approximation [3]. In this case, only the shear flow in the circumferential direction and the pressure flow in the axial direction are considered, the pressure flow in the circumferential direction being ignored. Whereas the oil film pressure is a function only of x in the case of an infinitely long bearing, it is a function of x and z in this case, namely the length (width) of bearing is taken into consideration, even though the bearing is assumed to be very short. In this case, mathematical handling is easy and it is recognized that the results coincide well with practice for small eccentricity ratios.

3.3.1 Oil Film Pressure

If the film thickness h is constant in the z direction, Eq. 3.52 becomes:

$$\frac{\partial^2 p}{\partial z^2} = \frac{6\mu U}{h^3} \frac{dh}{dx} \qquad (3.53)$$

Substituting the bearing clearance $h = c\,(1 + \kappa \cos \phi)$ into this yields:

$$\frac{\partial^2 p}{\partial z^2} = -\frac{6\mu U}{c^2 R} \frac{\kappa \sin \phi}{(1 + \kappa \cos \phi)^3} \qquad (3.54)$$

This can be integrated easily because the right-hand side is constant with respect to z. From symmetry, it can be assumed that the pressure gradient is zero, or $\partial p/\partial z = 0$, at the bearing center, $z = 0$. Then the pressure gradient will be:

$$\frac{\partial p}{\partial z} = -\frac{6\mu U}{c^2 R} \frac{\kappa \sin \phi}{(1 + \kappa \cos \phi)^3} z \qquad (3.55)$$

If this is integrated again under the assumption that the pressure is zero, or $p = 0$, at the bearing ends, $z = \pm(L/2)$, the pressure distribution is obtained as follows, L being the length (width) of the bearing metal:

$$p(\phi, z) = \frac{3\mu U}{c^2 R} \frac{\kappa \sin \phi}{(1 + \kappa \cos \phi)^3} \left(\frac{L^2}{4} - z^2 \right) \qquad (3.56)$$

It is seen in the above equation that the oil film pressure has a parabolic distribution in the z direction (the axial direction) and an antisymmetric distribution in the ϕ direction (the circumferential direction) with respect to the point ($\phi = \pi$, $p = 0$). The pressure distribution in the circumferential direction is similar to that in an infinitely long bearing.

3.3.2 Characteristics of a Short Bearing Under Gümbel's Condition

As a boundary condition for pressure, Gümbel's condition is used here once again, i.e., negative pressure in the region ($\pi \leq \phi \leq 2\pi$) will be replaced by zero.

a. Oil Film Force, Load Capacity, Eccentricity Ratio and Attitude Angle

The balance of the oil film force and the bearing load in the eccentricity direction and in the direction normal to it, respectively, can be written as follows:

$$2R \int_0^\pi \int_0^{L/2} p \cos \phi \, dz d\phi + P_1 \cos \theta = 0 \qquad (3.57)$$

$$2R \int_0^\pi \int_0^{L/2} p \sin \phi \, dz d\phi - P_1 \sin \theta = 0 \qquad (3.58)$$

By substituting Eq. 3.56 into the above equations and using the partial fraction decomposition method and J_3, J_2, and J_1 in Eq. 3.11, etc, the components of bearing load can be derived:

$$P_1 \cos \theta = \mu U R \left(\frac{R}{c}\right)^2 \left(\frac{L}{D}\right)^3 \frac{8\kappa^2}{(1-\kappa^2)^2} \qquad (3.59)$$

$$P_1 \sin \theta = \mu U R \left(\frac{R}{c}\right)^2 \left(\frac{L}{D}\right)^3 \frac{2\pi\kappa}{(1-\kappa^2)^{3/2}} \qquad (3.60)$$

From these equations, we have the bearing load as follows:

$$P_1 = 2\mu U R \left(\frac{R}{c}\right)^2 \left(\frac{L}{D}\right)^3 \frac{\kappa\{16\kappa^2 + \pi^2(1-\kappa^2)\}^{1/2}}{(1-\kappa^2)^2} \qquad (3.61)$$

This equation can be rewritten using the Sommerfeld number, $S = (\mu N/p_m)(R/c)^2$, as:

$$S \left(\frac{L}{D}\right)^2 = \frac{(1-\kappa^2)^2}{\pi\kappa\{16\kappa^2 + \pi^2(1-\kappa^2)\}^{1/2}} \qquad (3.62)$$

Further, from Eq. 3.59 and Eq. 3.60, we can write the locus of the bearing center as:

$$\tan \theta = \frac{\pi(1-\kappa^2)^{1/2}}{4\kappa} \qquad (3.63)$$

This is shown in Fig. 3.13. The locus has the form of a half-moon that is a little thinner than that of Fig. 3.11.

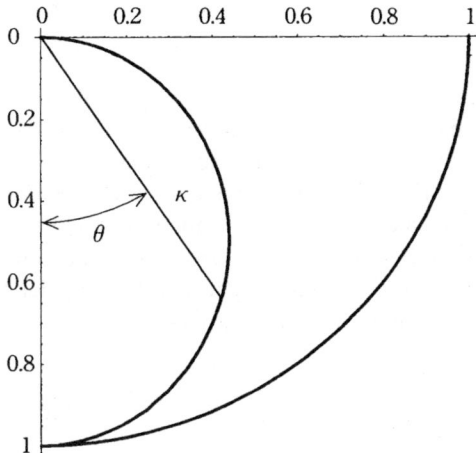

Fig. 3.13. Locus of the bearing center for a short bearing under Gümbel's condition

b. Frictional Moment

Since the pressure flow in the circumferential direction is disregarded and $(\partial p/\partial \phi) = 0$ is assumed in the short bearing approximation, the frictional moment acting on the journal and the bearing metal are the same. If they are denoted by M, it can be calculated as follows by using J_1, as before:

$$M = \int_0^{2\pi} \frac{\mu U}{h} LR^2 d\phi = \frac{2\pi \mu U LR^2}{c} \frac{1}{(1-\kappa^2)^{1/2}} \quad (3.64)$$

c. Leakage of Lubricating Oil

While leakage of oil from the end of a bearing cannot be considered in an infinitely long bearing, it is calculable in the case of a short bearing. The flow rate q of leaking oil from a bearing end can be obtained by using Eq. 3.56 as follows:

$$q = 2\int_0^h dy \int_0^\pi \frac{h^3}{12\mu}\left[\frac{dp}{dz}\right]_{z=L/2} R\,d\phi = ULc\kappa \quad (3.65)$$

The influence of oil supply pressure is not considered here.

3.4 Finite Length Bearings

In the case of a finite length bearing, it is necessary to solve the following two-dimensional Reynolds' equation:

$$\frac{\partial}{\partial x}\left(h^3 \frac{\partial p}{\partial x}\right) + \frac{\partial}{\partial z}\left(h^3 \frac{\partial p}{\partial z}\right) = 6\mu U \frac{\partial h}{\partial x} \qquad (3.66)$$

where $U = U_2$. However, the above equation cannot be solved analytically in general, but must be solved numerically. Numerical solutions include the orthogonal function expansion method [4] [12], the finite element method [9] [10] [11] and the finite difference method [13]. The finite difference method will be described here.

For the sake of convenience, the above equation is first nondimensionalized by the introduction of the following nondimensional quantities in which D, L, and ω are the bearing diameter, the bearing length, and the angular velocity of rotation, respectively. The other symbols are the same as before:

$$\bar{x} = \frac{x}{D}, \quad \bar{z} = \frac{z}{L}, \quad \bar{h} = \frac{h}{c}, \quad \bar{p} = \frac{p}{\mu\omega}\left(\frac{c}{R}\right)^2 \qquad (3.67)$$

Then, Reynolds' equation, Eq. 3.66 can be nondimensionalized in the following form, where $U = R\omega$:

$$\frac{\partial}{\partial \bar{x}}\left(\bar{h}^3 \frac{\partial \bar{p}}{\partial \bar{x}}\right) + \left(\frac{D}{L}\right)^2 \frac{\partial}{\partial \bar{z}}\left(\bar{h}^3 \frac{\partial \bar{p}}{\partial \bar{z}}\right) = 12 \frac{\partial \bar{h}}{\partial \bar{x}} \qquad (3.68)$$

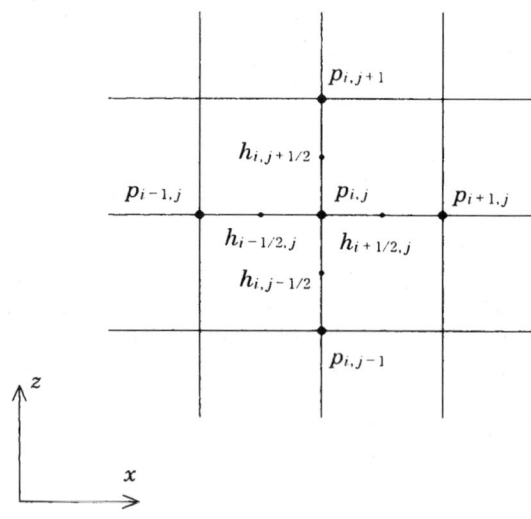

Fig. 3.14. Grid on a lubricating surface

The lubricating domain is divided into a grid pattern as shown in Fig. 3.14 and then the three derivatives in the above equation are discretized on the grid as follows:

$$\frac{\partial}{\partial x}\left(h^3\frac{\partial p}{\partial x}\right) = \frac{h_{i+1/2,j}^3\frac{p_{i+1,j}-p_{i,j}}{\Delta x} - h_{i-1/2,j}^3\frac{p_{i,j}-p_{i-1,j}}{\Delta x}}{\Delta x} \qquad (3.69)$$

$$\frac{\partial}{\partial z}\left(h^3\frac{\partial p}{\partial z}\right) = \frac{h_{i,j+1/2}^3\frac{p_{i,j+1}-p_{i,j}}{\Delta z} - h_{i,j-1/2}^3\frac{p_{i,j}-p_{i,j-1}}{\Delta z}}{\Delta z} \qquad (3.70)$$

$$\frac{\partial h}{\partial x} = \frac{h_{i+1/2,j} - h_{i-1/2,j}}{\Delta x} \qquad (3.71)$$

Substituting these into Eq. 3.68 and solving for $p_{i,j}$, we obtain $p_{i,j}$ in the following form:

$$p_{i,j} = a_0 + a_1 p_{i+1,j} + a_2 p_{i-1,j} + a_3 p_{i,j+1} + a_4 p_{i,j-1} \qquad (3.72)$$
$$(i = 1, 2, \cdots, m-1; \; j = 1, 2, \cdots, n-1)$$

where $p_{i,j}$ is the pressure at the nodal point (i, j) and a_0, a_1, a_2, a_3, and a_4 are constants given at the respective nodal point.

If the number of nodal points at which the pressure is to be calculated is N, then we have N equations of the form of Eq. 3.72. The boundary conditions are given as the pressure at the points on the boundary. By solving these simultaneously, the pressures at respective nodal points are obtained.

One of the methods to solve these simultaneous equations is **elimination**, typically Gauss' elimination method. Pressure at all the nodal points can thereby be found in a finite number of operations.

Further, if we sweep the lubricating domain two dimensionally with the calculation of Eq. 3.72 at each nodal point in consecutive order starting from a suitable nodal point, and if this is repeated a sufficient number of times, then it is expected that the pressure obtained at each nodal point gradually approaches the true value of the pressure. This is called the **iterative method** (successive approximation method). In this case, the calculation will be repeated until the following relation is satisfied:

$$\frac{\sum_{i,j}|(p_{i,j})_k - (p_{i,j})_{k-1}|}{\sum_{i,j}|(p_{i,j})_k|} < \epsilon \qquad (3.73)$$

where $(p_{i,j})_k$ are pressures obtained in the kth calculation, $(p_{i,j})_{k-1}$ are those in the previous calculation, and ϵ is a sufficiently small allowable error. The pressure $(p_{i,j})_k$ obtained in the kth calculation will be a solution $p_{i,j}$. For a high accuracy of calculation, a value of $10^{-6} - 10^{-12}$, for example, is used for ϵ. Further, the convergence in the region of rapid pressure rise can be improved by introducing the following new variable $\bar{p}'$ into Reynolds' equation Eq. 3.68 [2]. $\bar{h}$ and $\bar{p}$ are the nondimensional film thickness and nondimensional pressure in Eq. 3.67, respectively:

$$\bar{p}' = \bar{h}^m \bar{p} \qquad (3.74)$$

A value of $m = 2$ or 1.5, for example, is used [6][13].

If pressure $p_{i,j}$ is obtained, the component of the oil film force P in the eccentricity direction and that in the direction normal to it, respectively, are calculated as follows:

$$P_\kappa = P \cos \theta = R \sum_{i,j} p_{i,j} \cos \phi_i \Delta z \Delta \phi \qquad (3.75)$$

$$P_\theta = P \sin \theta = R \sum_{i,j} p_{i,j} \sin \phi_i \Delta z \Delta \phi \qquad (3.76)$$

where $\Delta x = R \Delta \phi$.

References

1. A. Sommerfeld, "Zur hydrodynamischen Theorie der Schmiermittelreibung", *Zeit. angew. Math. u. Physik*, Vol. 50, 1904, pp. 97 - 155.
2. G. Vogelpohl, "Beiträge zur Kenntniss der Gleitlagerreibung", *VDI - Forschungsheft* 386, Berlin, 1937.
3. G.B. DuBois and F.W. Ocvirk, "Analytical Derivation and Experimental Evaluation of Short Bearing Approximation for Full Journal Bearings", NACA Tech. Report 1157, Washington D.C., 1953
4. T. Someya, "Stabilität einer in zylindrischen Gleitlagern laufenden, unwuchtfreien Welle", *Ingenieur-Archiv*, 33. Band, 2. Heft, 1963, pp. 85 - 108.
5. N. Soda, "Bearings" (in Japanese), *Iwanami Zensho Series, Iwanami Shoten*, Tokyo, 1964.
6. T. Someya, "Das dynamisch belastete Radial-Gleitlager beliebigen Querschnitts", *Ingenieur-Archiv*, 34. Band, 1. Heft, 1965, pp. 7 - 16.
7. H. Mori, S. Miyata, Y. Abe and Y. Fujita, "Research on Discountinuity of Lubricant Film in Journal Bearing (First Report, Bearing Characteristics and Oil Film Rupture)" (in Japanese), *Trans JSME*, Vol. 33, No. 248, April 1967, pp. 658 - 666.
8. H. Mori, H. Yabe, Y. Fujita, Y. Iio and Y. Okumoto, "Ditto (Second Report, Separation Boundary Condition)" (in Japanese), *Trans. JSME*, Vol. 35, No. 272, April 1969, pp. 899 - 906.
9. M.M. Reddi, "Finite Element Solution of the Incompressible Lubrication Problems", *Trans. ASME. F*, Vol. 91, No. 3, July 1969, pp. 524 - 533.
10. S. Wada, H. Hayashi and M. Migita, "Application of Finite Element Method to Hydrodynamic Lubrication Problems (Part 1)" (in Japanese), *Trans. JSME*, Vol. 37, No. 295, March 1971, pp. 583 - 592.
11. S. Wada and H. Hayashi, "Application of Finite Element Method to Hydrodynamic Lubrication Problems (Part 2)" (in Japanese), *Trans. JSME*, Vol. 37, No. 295, March 1971, pp. 593 - 601.
12. T. Kato and Y. Hori, "Matrix Form of Reynolds' Equation", *JSME International Journal*, Vol. 31, No. 2, June 1986, pp. 444 - 450.
13. J. Mitsui, "Method of Calculation for Bearing Characteristics", Journal Bearing Databook (T. Someya, Editor), *Springer Verlag*, Berlin Heidelberg, 1988, pp. 231 - 240

4
Fundamentals of Thrust Bearings

An axial load acting on a shaft is called a thrust. A bearing that supports a thrust is called a thrust bearing. In this chapter, the fundamentals of thrust bearings are discussed.

In a journal bearing, an oil film wedge is automatically formed due to journal eccentricity because of the bearing load; therefore, a load capacity is automatically generated by the load itself. In a thrust bearing, however, an oil film wedge is not formed automatically, so it must be prepared artificially to support the load. This is an essential difference between a journal bearing and a thrust bearing.

In the early days, a thrust bearing consisted of a rotating disk fixed to a shaft and a mating stationary disk (the shaded part) as shown in Fig. 4.1a-ca. Since these disks were parallel to each other, an oil film wedge was not formed between them, and hence the load capacity was theoretically zero. Therefore, solid friction took place between the disks and so severe heat generation and wear occurred. This was a big problem in the bearings of propeller shafts for ships and in the bearings of vertical shafts for water turbines, for example.

One attempt to overcome these problems was to form an oil film wedge artificially between a rotating disk fixed to the shaft and a spatially fixed inclined pad (the shaded wedge-shaped pad) as shown in Fig. 4.1a-cb. This seems at first to be a good idea, but since the optimal angle of inclination of the pad is very small, it is difficult to finish the inclined surface with sufficient accuracy. Furthermore, even if it could be successfully manufactured, the inclination may change afterward as a result of elastic deformation or thermal deformation, for example.

A solution to this problem was found independently by A. G. M. Michell (1870–1959) of Australia and A. Kingsbury (1863–1943) of the USA. Their idea was to support an inclined plate at a certain point a little downstream of the center by a pivot so that it can tilt freely. It is shown schematically in Fig. 4.1a-cc. Since a pivoted plate (or a pad) has the property that its inclination is automatically determined by the pivot position, an oil film wedge of optimal inclination can be formed automatically by choosing a suitable pivot position. This approach resulted in the first reasonable design for a thrust bearing.

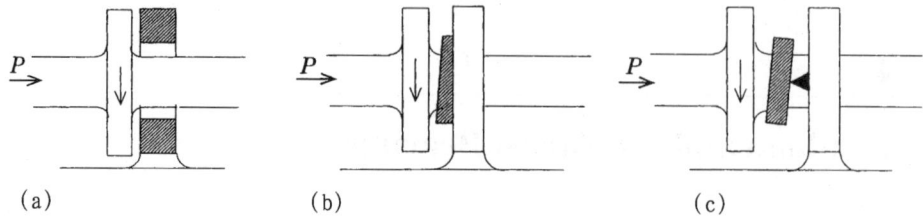

Fig. 4.1a-c. Development of thrust bearings. **a** parallel disks, **b** fixed inclined pad, **c** tilting pad. The shaded patern is the stationary disk

Such a bearing is called a **Michell bearing** or a **Kingsbury bearing** . It is also called a **tilting pad bearing** from its working principle. Although Michell obtained a patent for this type of bearing in 1905 [1] and Kingsbury also obtained a patent for the same idea in 1910 [3], it is known that Kingsbury materialized his bearing before Michell's acquisition of the patent, and it is recognized that their inventions were independent. As for the pronunciation of Michell's name, according to D. Dowson [23], Michell himself explained that his family name was pronounced with emphasis on the first syllable and that it should rhyme with "rich."

It is interesting indeed that hydrodynamic lubrication theory was developed from the investigation into journal bearings, and thrust bearings were invented in turn as an application of lubrication theory.

4.1 Infinitely Long Plane Pad Bearings

As shown in Fig. 4.2, a plane which moves with a velocity U and a plane pad with a slight inclination to the moving plane are considered. For simplicity, it is assumed that the pad is infinitely long in the direction normal to the page.

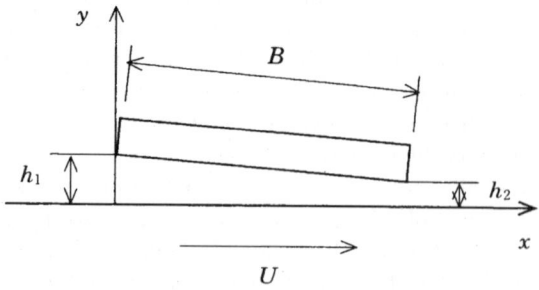

Fig. 4.2. Plane pad bearing

4.1.1 Basic Formulae

The thickness h of an oil film between the two surfaces in Fig. 4.2 is expressed by a linear equation of the coordinate x as follows:

$$h(x) = h_1 - \frac{x}{B}(h_1 - h_2) \tag{4.1}$$

where h_1 and h_2 are the film thickness at the entrance and the exit of the pad, respectively, and B is the width of the pad in the direction of the plane motion.

If nondimensional quantities

$$m = \frac{h_1}{h_2}, \qquad \bar{x} = \frac{x}{B}, \qquad \bar{h} = \frac{h}{h_2} \tag{4.2}$$

are introduced, Eq. 4.1 can be written as follows:

$$\bar{h}(x) = m - \bar{x}(m - 1) \tag{4.3}$$

where m is a parameter indicating the inclination of the pad.

Reynolds' equation for the pad can be written as follows, the pad being assumed to be infinitely long in the direction normal to the page:

$$\frac{d}{dx}\left(h^3 \frac{dp}{dx}\right) = 6\mu U \frac{dh}{dx} \tag{4.4}$$

where p is the pressure, μ is the coefficient of viscosity, and U is the velocity of the moving plane.

In the case of a thrust bearing, the boundary conditions for pressure are usually simple. Namely, pressure p can be set at zero at the entrance and the exit of the bearing. In the case of a thrust bearing, film thickness does not usually increase in the downstream direction at any point and so oil film rupture will not occur in the film as it does in the case of a journal bearing.

4.1.2 Basic Characteristics

a. Pressure Distribution

Integrating Eq. 4.4 with respect to x gives:

$$\frac{dp}{dx} = 6\mu U \left(\frac{1}{h^2} + \frac{C_1}{h^3}\right)$$

where C_1 is the integral constant. If the oil film thickness at the point of maximum pressure, or at the point where $dp/dx = 0$, is denoted by h_m, we have:

$$C_1 = -h_m$$

from the above equation, thus:

$$\frac{dp}{dx} = 6\mu U \left(\frac{1}{h^2} - \frac{h_m}{h^3}\right) \quad (4.5)$$

where h_m is still unknown. If this is integrated again, then:

$$p = 6\mu U \int_{h_1}^{h} \left(\frac{1}{h^2} - \frac{h_m}{h^3}\right) dx + C_2 \quad (4.6)$$

where h_m and C_2 are integral constants that can be determined by the boundary conditions, i.e., that the pressure is zero at the entrance and the exit, i.e.,

$$p(h_1) = p(h_2) = 0 \quad (4.7)$$

Let us first consider the entrance, where $h = h_1$. From Eq. 4.6 and Eq. 4.7, we have:

$$p(h_1) = 0 + C_2 = 0, \quad \text{therefore} \quad C_2 = 0 \quad (4.8)$$

Next, consider the exit. Since $h = h_2$ there, we have:

$$p(h_2) = 6\mu U \int_{h_1}^{h_2} \left(\frac{1}{h^2} - \frac{h_m}{h^3}\right) dx = 0$$

Then h_m will be determined as follows:

$$h_m = \left(\int_{h_1}^{h_2} \frac{1}{h^2} dx\right) \Big/ \left(\int_{h_1}^{h_2} \frac{1}{h^3} dx\right) \quad (4.9)$$

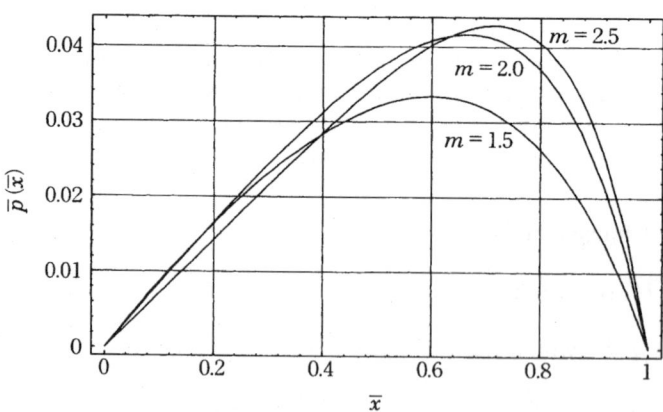

Fig. 4.3. Nondimensional pressure distribution in an infinitely long plane pad bearing

The integral constants of Eq. 4.6 are now determined, hence the pressure distribution p is determined. It can be written as follows if nondimensional variables $\bar{h}$ and $\bar{x}$ are used:

$$p = \frac{6\mu UB}{h_2{}^2} \int_{h_1}^{h} \left(\frac{1}{\bar{h}^2} - \frac{\bar{h}_m}{\bar{h}^3} \right) d\bar{x} \tag{4.10}$$

where
$$\bar{h}_m = \frac{h_m}{h_2} = \left(\int_{h_1}^{h_2} \frac{1}{\bar{h}^2} dx \right) \Big/ \left(\int_{h_1}^{h_2} \frac{1}{\bar{h}^3} dx \right) \tag{4.11}$$

Since $\bar{h}$ is given by Eq. 4.3 in the case of a plane pad bearing, $p(\bar{x})$ and $\bar{h}_m$ will be as follows:

$$p(\bar{x}) = \frac{6\mu UB}{h_2{}^2} \frac{(m-1)(1-\bar{x})\bar{x}}{(m+1)(m-m\bar{x}+\bar{x})^2} \equiv \frac{6\mu UB}{h_2{}^2} \bar{p}(\bar{x}) \tag{4.12}$$

$$\bar{h}_m = \frac{2m}{m+1} \tag{4.13}$$

Nondimensional pressure $\bar{p}(\bar{x})$ on the right-hand side of Eq. 4.12 is shown in Fig. 4.3 with m as a parameter. This shows the shape of pressure distribution.

b. Load Capacity

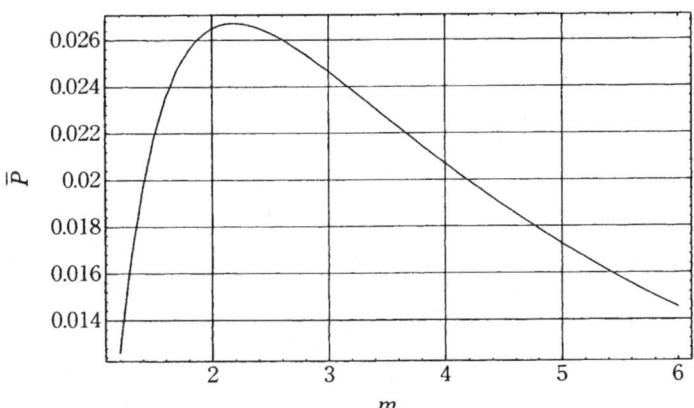

Fig. 4.4. Nondimensional load capacity of an infinitely long plane pad bearing as a function of pad inclination

Integrating the oil film pressure over the pad width gives the load capacity of the pad bearing. Load capacity P per unit length will be:

$$P = \int_{h_1}^{h_2} p\, dx = \left[xp \right]_{h_1}^{h_2} - \int_{h_1}^{h_2} x \frac{dp}{dx} dx = -\int_{h_1}^{h_2} x \frac{dp}{dx} dx \tag{4.14}$$

$$= 6\mu U \int_{h_1}^{h_2} x \left(\frac{1}{h^2} - \frac{h_m}{h^3} \right) dx = \frac{6\mu UB^2}{h_2{}^2} \int_{h_1}^{h_2} \bar{x} \left(\frac{1}{\bar{h}^2} - \frac{\bar{h}_m}{\bar{h}^3} \right) d\bar{x} \tag{4.15}$$

In the case of an infinitely long plane pad bearing, the load capacity can be obtained as follows by using Eq. 4.12:

$$P = \frac{6\mu U B^2}{h_2^2} \frac{1}{(m-1)^2} \left[\ln m - \frac{2(m-1)}{m+1} \right] \equiv \frac{6\mu U B^2}{h_2^2} \bar{P}(m) \qquad (4.16)$$

The relation between the nondimensional load capacity $\bar{P}(m)$ and the pad inclination m is shown in Fig. 4.4. The figure shows that the nondimensional load capacity has a maximum in the neighborhood of a pad inclination of $m = 2.2$.

c. Center of Pressure and Pivot Position

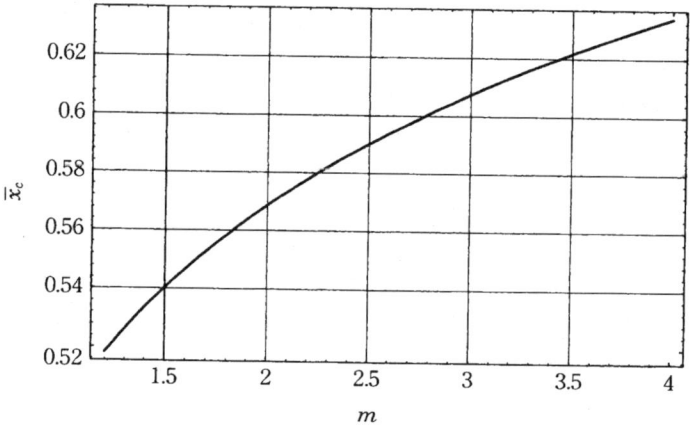

Fig. 4.5. Center of pressure (i.e., the pivot postion) of an infinitely long plane pad as a function of pad inclination

The position x_c of the center of oil film pressure is given as follows:

$$x_c = \frac{1}{P} \int_{h_1}^{h_2} p\, x\, dx = \frac{3\mu U B^3}{P h_2^2} \int_{h_1}^{h_2} \bar{x}^2 \left(\frac{1}{\bar{h}^2} - \frac{\bar{h}_m}{\bar{h}^3} \right) d\bar{x} \qquad (4.17)$$

In the case of an infinitely long plane pad bearing, the nondimensional center of pressure, $\bar{x}_c = x_c/B$, is given as follows for a given inclination m, by using Eq. 4.12:

$$\bar{x}_c = \frac{2m(m+2)\ln m - (m-1)(5m+1)}{2(m-1)\{(1+m)\ln m - 2(m-1)\}} \qquad (4.18)$$

The dependence of $\bar{x}_c$ on m is shown in Fig. 4.5. It is important to note that $\bar{x}_c$ increases monotonously with m.

This figure can also be interpreted as a graph that gives the inclination of a pad for a given pivot position. If a pad is supported by a pivot at a certain position, the position must coincide with the center of pressure $\bar{x}_c$ from a balance of the moments acting on the pad. Therefore, the pad takes an inclination m corresponding to $\bar{x}_c$ automatically. For example, if the pivot position is taken as $\bar{x} = 0.57$, it must be at the center of pressure $\bar{x}_c$ and so, from the figure, the inclination will be $m = 2$. This is a value which is automatically determined. If the inclination m is larger than this value, the center of pressure $\bar{x}_c$ will be located downstream of the pivot position, and hence the pad is subject to a moment which makes its inclination smaller. If the inclination m is smaller than the value, the pad is subject to a moment which makes its inclination larger.

More precisely, if the pad is supported by a pivot at the position of $\bar{x}_c = 0.578$ (the position 57.8% along the pad from the entrance of the pad), the inclination of the pad becomes $m = 2.2$ automatically, and then, according to Fig. 4.4, the load capacity will be maximum. This is the working principle of a Michell bearing or a Kingsbury bearing.

d. Frictional Force

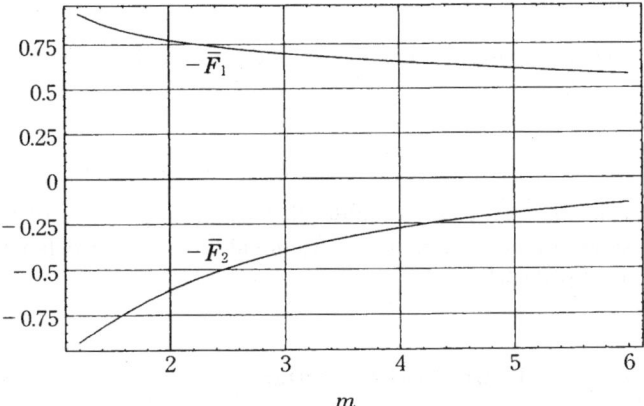

Fig. 4.6. Nondimensional frictional force of an infinitely long plane pad as a function of pad inclination. $\bar{F}_1$ is the force on the moving surface and $\bar{F}_2$ is the force on the fixed pad

The shear stresses in the oil film at a moving surface and a stationary pad surface are obtained as follows with $U_1 = U$ and $U_2 = 0$, similarly to Eq. 3.33 and Eq. 3.34:

$$\tau_{y=0} = -\frac{\mu U}{h} - \frac{h}{2}\frac{dp}{dx} \qquad (4.19)$$

$$\tau_{y=h} = -\frac{\mu U}{h} + \frac{h}{2}\frac{dp}{dx} \qquad (4.20)$$

The moving surface is considered first. Integrating Eq. 4.19, with Eq. 4.5 substituted into it, over the width of the plate gives a frictional force acting on the moving surface per unit length as follows:

$$F_1 = \mu U \int_{h_1}^{h_2} \left(\frac{3h_m}{h^2} - \frac{4}{h} \right) dx \tag{4.21}$$

$$= \frac{\mu U B}{h_2} \int_{h_1}^{h_2} \left(\frac{3\bar{h}_m}{\bar{h}^2} - \frac{4}{\bar{h}} \right) d\bar{x} \tag{4.22}$$

In the case of an infinitely long plane pad, substituting Eq. 4.3 and Eq. 4.13 into the above equation yields the following frictional force acting on the moving surface:

$$F_1 = \frac{\mu U B}{h_2} \frac{1}{m-1} \left[-4 \ln m + \frac{6(m-1)}{m+1} \right] \equiv \frac{\mu U B}{h_2} \bar{F}_1(m) \tag{4.23}$$

Similarly, from Eq. 4.20, the frictional force acting on the fixed pad surface is obtained as follows:

$$F_2 = \frac{\mu U B}{h_2} \frac{1}{m-1} \left[2 \ln m - \frac{6(m-1)}{m+1} \right] \equiv \frac{\mu U B}{h_2} \bar{F}_2(m) \tag{4.24}$$

Nondimensional frictional forces $\bar{F}_1(m)$ and $\bar{F}_2(m)$ are shown in Fig. 4.6 against the pad inclination m.

As is seen in the figure, F_1 and F_2 are not equal. Calculating their difference leads to:

$$F_2 - F_1 = \frac{h_1 - h_2}{B} P \tag{4.25}$$

which is equal to the load capacity multiplied by the inclination of the fixed pad surface. This shows that the difference of frictional forces is attributable to the inclination of the fixed pad surface.

4.2 Finite Length Plane Pad Bearings

If the length of a pad (the length in the direction normal to the page) is finite, the load capacity falls markedly because of leakage of lubricating oil from both ends of the pad. The ratio of the load capacity per unit length of a finite length pad bearing to that of an infinitely long pad bearing is called the side leakage factor.

A rigorous analysis of a finite length plane pad bearing was performed for the first time by Michell [2]. He expressed the pressure distribution p in the length direction (the z direction) as an infinite series of sin functions of odd terms as follows:

$$p = p_1 + p_3 + p_5 + \cdots + p_m + \cdots \tag{4.26}$$

where $p_m = \dfrac{w_m(x) \sin mz}{mx}$, m is a positive odd number where $w_m(x)$ are functions of x only. Substituting Eq. 4.26 in the left-hand side of Reynolds' equation, expressing the right-hand side also with an infinite series of sin functions of odd terms, and letting the coefficients of like terms on the right and left sides be equal, the differential equations for $w_m(x)$ can be obtained.

Major developments were achieved by Michell not only in the invention of the Michell bearing but also in the theory of hydrodynamic lubrication [4].

Nowadays, analyses of finite length pad bearings are usually performed by numerical computation, by means of the finite difference method or the finite element method, for example.

4.3 Sector Pad Bearings

In an actual thrust bearing, four to seven sector pads such as that shown in Fig. 4.7 are usually used. The pads are arranged in a circular form, facing the rotating disk. So far, Reynolds' equation has been considered in rectangular coordinates. It is natural, however, to deal with a sector pad in cylindrical coordinates. Therefore, Reynolds' equation in cylindrical coordinates is considered here. The dot in the pad in the figure shows the pivot position.

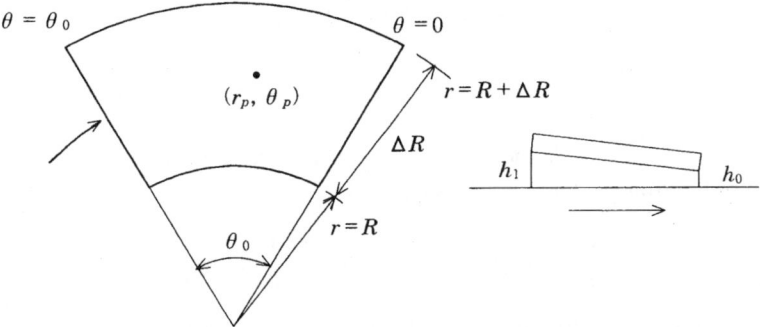

Fig. 4.7. A sector pad. The *dot* shows the pivot postion

4.3.1 Reynolds' Equation in Cylindrical Coordinates

A stationary flow of incompressible viscous fluid is considered. The balance of forces acting on a small volume element in cylindrical coordinates (r, θ, z) is expressed as follows:

$$\mu \frac{\partial^2 v_r}{\partial z^2} = \frac{\partial p}{\partial r} - \frac{\rho v_\theta^2}{r} \tag{4.27}$$

4 Fundamentals of Thrust Bearings

$$\mu \frac{\partial^2 v_\theta}{\partial z^2} = \frac{1}{r} \frac{\partial p}{\partial \theta} \tag{4.28}$$

$$\frac{\partial p}{\partial z} = 0 \tag{4.29}$$

where v_r and v_θ are the flow velocities in the radial and the circumferential directions, respectively, and the second term on the right-hand side of Eq. 4.27 indicates the centrifugal force.

Integrating Eq. 4.28 under the boundary conditions $v_\theta = r\omega$ at $z = 0$, $v_\theta = 0$ at $z = h$ yields:

$$v_\theta = \frac{1}{2\mu r} \frac{\partial p}{\partial \theta} z(z-h) - \frac{r\omega}{h}(z-h) \tag{4.30}$$

where ω is the angular velocity of the rotating disk. Substituting this into Eq. 4.27 and integrating it under the boundary conditions $v_r = 0$ at $z = 0$ and $z = h$ gives:

$$v_r = \frac{1}{2\mu} \frac{\partial p}{\partial r} z(z-h) + \frac{\rho}{\mu r}\left(\frac{z}{h} \int_0^h \int_0^z (v_\theta^2) dz dz - \int_0^z \int_0^z (v_\theta^2) dz dz\right) \tag{4.31}$$

The continuity equation for an incompressible fluid in cylindrical coordinates can be written as follows:

$$\frac{\partial (r v_r)}{\partial r} + \frac{\partial v_\theta}{\partial \theta} + r \frac{\partial v_z}{\partial z} = 0 \tag{4.32}$$

where v_z is the flow velocity in the oil film thickness direction. Integrating the above equation with respect to z from 0 to h under the boundary conditions $v_z = 0$ at $z = 0$, $v_r = v_\theta = v_z = 0$ at $z = h$ and using the mathematical formula for exchange of the order of differentiation and integration gives:

$$r \frac{\partial}{\partial r} \int_0^h v_r \, dz + \int_0^h v_r \, dz + \frac{\partial}{\partial \theta} \int_0^h v_\theta \, dz = 0 \tag{4.33}$$

Substituting Eq. 4.30 and Eq. 4.31 into Eq. 4.33 yields Reynolds' equation in cylindrical coordinates as follows:

$$\frac{\partial}{\partial r}\left(\frac{h^3}{12\mu} \frac{\partial p}{\partial r} + G_c\right) + \frac{1}{r}\left(\frac{h^3}{12\mu} \frac{\partial p}{\partial r} + G_c\right) + \frac{1}{r^2} \frac{\partial}{\partial \theta}\left(\frac{h^3}{12\mu} \frac{\partial p}{\partial \theta}\right) = \frac{\omega}{2} \frac{\partial h}{\partial \theta} \tag{4.34}$$

where

$$G_c = \frac{\rho}{\mu r} \int_0^h \left(\frac{z}{h} \int_0^h \int_0^z (v_\theta^2) dz dz - \int_0^z \int_0^z (v_\theta^2) dz dz\right) dz \tag{4.35}$$

G_c is the integration of the second half of the right-hand side of Eq. 4.31 with respect to z and is a term related to the centrifugal force. Also, if the sign of the right-hand side of Eq. 4.34 is considered in the case of Fig. 4.7, it is seen that (the right-hand side) < 0 because $\omega < 0$, since the disk is rotating in the clockwise direction, and $\partial h/\partial \theta > 0$. Therefore, a convex pressure distribution and hence a positive pressure develops. The same is obtained also in the case of $\omega > 0$ and $\partial h/\partial \theta < 0$.

If it is assumed, for simplicity, that the pressure gradient in Eq. 4.30 can be disregarded, v_θ will be:

$$v_\theta = -\frac{r\omega}{h}(z-h) \tag{4.36}$$

and then it will be seen that v_r and G_c are given as follows, respectively, from Eq. 4.31 and Eq. 4.35 [24]:

$$v_r = \frac{1}{2\mu}\frac{\partial p}{\partial r}z(z-h) + \frac{\rho r \omega^2}{\mu}\left(\frac{hz}{4} - \frac{z^2}{2} + \frac{z^3}{3h} - \frac{z^4}{12h^2}\right) \tag{4.37}$$

$$G_c = \frac{\rho r \omega^2}{\mu}\frac{h^3}{40} \tag{4.38}$$

If $G_c = 0$ is assumed in Eq. 4.34, Reynolds' equation ignoring the centrifugal force will be as follows:

$$\frac{\partial}{\partial r}\left(h^3\frac{\partial p}{\partial r}\right) + \frac{h^3}{r}\frac{\partial p}{\partial r} + \frac{1}{r^2}\frac{\partial}{\partial \theta}\left(h^3\frac{\partial p}{\partial \theta}\right) = 6\mu\omega\frac{\partial h}{\partial \theta} \tag{4.39}$$

or

$$\frac{\partial}{\partial r}\left(rh^3\frac{\partial p}{\partial r}\right) + \frac{1}{r}\frac{\partial}{\partial \theta}\left(h^3\frac{\partial p}{\partial \theta}\right) = 6\mu r \omega \frac{\partial h}{\partial \theta} \tag{4.40}$$

These equations can also be derived from Reynolds' equation in rectangular coordinates by the following coordinate transformation:

$$x = r\cos\theta, \quad z = -r\sin\theta \tag{4.41}$$

4.3.2 Numerical Solution of a Sector Pad

Let us solve the sector pad problem numerically using Eq. 4.39. Namely:

$$\frac{\partial}{\partial r}\left(h^3\frac{\partial p}{\partial r}\right) + \frac{h^3}{r}\frac{\partial p}{\partial r} + \frac{1}{r^2}\frac{\partial}{\partial \theta}\left(h^3\frac{\partial p}{\partial \theta}\right) = 6\mu\omega\frac{\partial h}{\partial \theta} \tag{4.42}$$

The following nondimensional quantities are introduced here using the inner radius R, the width in the radial direction ΔR, and the angular extent in the circumferential direction θ_0 of the sector pad; the angular velocity ω of the disk; and the exit clearance h_0 of the pad (see Fig. 4.7).

$$\bar{r} = \frac{r-R}{\Delta R}, \quad \bar{\theta} = \frac{\theta}{\theta_0}, \quad \bar{h} = \frac{h}{h_0}, \quad \bar{p} = \frac{p h_0^2}{\omega \mu R^2} \tag{4.43}$$

4 Fundamentals of Thrust Bearings

Equation 4.42 is nondimensionalized as follows with the above nondimensional quantities:

$$\frac{\partial}{\partial \bar{r}}\left(\bar{h}^3 \frac{\partial \bar{p}}{\partial \bar{r}}\right) + \frac{\Delta R}{R + \bar{r}\Delta R} \bar{h}^3 \frac{\partial \bar{p}}{\partial \bar{r}} + \left(\frac{\Delta R}{R + \bar{r}\Delta R}\right)^2 \frac{1}{\theta_0^2} \frac{\partial}{\partial \bar{\theta}}\left(\bar{h}^3 \frac{\partial \bar{p}}{\partial \bar{\theta}}\right)$$
$$= 6\left(\frac{\Delta R}{R}\right)^2 \frac{1}{\theta_0} \frac{\partial \bar{h}}{\partial \bar{\theta}} \quad (4.44)$$

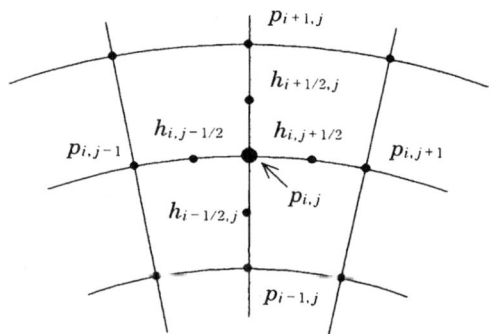

Fig. 4.8. Grid for a sector pad

Next, the sector pad is divided into a grid as shown in Fig. 4.8. If the differential coefficients in the above equation are expressed by the finite quantities of the grid and substituted into the above equation, the pressure $p_{i,j}$ at an arbitrary nodal point can be written in the following form:

$$p_{i,j} = a_0 + a_1 p_{i+1,j} + a_2 p_{i-1,j} + a_3 p_{i,j+1} + a_4 p_{i,j-1} \quad (4.45)$$

With this expression and the boundary conditions, the pressure at all the nodal points can be obtained by the method of successive approximation or elimination.

4.4 Additional Topics

4.4.1 Influence of Deformation of the Pad

In this chapter, it has been assumed that the lubricating surface of the pad always remains flat, without any deformation. However, the actual surface of the pad may warp due to the pivot support or the heat generation in the lubricating surface. In both the cases, the lubricating surface becomes convex. Such a deformation naturally changes the lubrication characteristics of the pad.

In the design of a pad, it is important to reduce the elastic deformation of the pad by making it sufficiently thick, by distributing the supporting points, or by reducing the thermal deformation of the pad by suitable heat removal or heat isolation [4] [7] [8] [9] [10].

4.4.2 Magnetic Disk Memory Storage

In magnetic disk memory devices widely used in computers in recent years, a read/write element is attached to a slider which floats over a rotating magnetic disk surface supported by a very thin air film [6]. In using such a mechanism, it is expected that the slider will trace the oscillatory motion of the disk surface during rotation in such a way that the gap between the read/write element and the disk surface is kept constant. In fact, however, traceability of the slider is a big problem [11][13][19][20]. The slider in this case floats on an air film on the same principle as that of the pad of a thrust bearing, the lubricating air film being automatically formed using the surrounding air as lubricant.

In the case of magnetic disk memory devices, a small gap between the read/write element and the disk surface is required for a high density of recording. Therefore, efforts have been made to make the floating height of a slider as small as possible. The air film thickness h has become as small as 10–20 nm in recent years. This means that h is of the same order of, or smaller than, the mean free path λ (approximately 70 nm) of air molecules. Or, unlike in usual engineering problems, the Knudsen number $M = \lambda/h$ is not very small, but is of the same order of, or in some cases significantly larger than, one.

This means that the air cannot be treated as a continuum, but must be treated as an ensemble of particles. The effect of the particulate nature appears as slip between the wall surface and air (slip flow), or equivalently, appears as a decrease in the viscosity of the air. A modified Reynolds' equation considering this effect was derived by Burgdorfer as follows [5].

$$\frac{\partial}{\partial x}\left[ph^3\frac{\partial p}{\partial x}\left(1+6\frac{\lambda}{h}\right)\right] + \frac{\partial}{\partial y}\left[ph^3\frac{\partial p}{\partial y}\left(1+6\frac{\lambda}{h}\right)\right] = 6\mu U\frac{\partial}{\partial x}(ph) \qquad (4.46)$$

This equation was derived for an air journal bearing, and is said to be applicable when $0 < M \ll 1$. In recent magnetic disk memory devices, the air film thickness is very small and the above condition for M does not seem to be satisfied. It is reported, however, that Eq. 4.46 is in fact applicable down to the level of $M \approx 1$ (or $h = 100$ nm $= 0.1$ μm) [12][14]. Analyses of the floating characteristics of sliders in such cases are carried out by using this equation [13][19][20][22]. In an experiment on a centrally supported catamaran type slider consisting of two convex shoes 5.8 mm long, 1.8 mm wide, with a 2-μm swell at the center, it is reported that the measured floating height was 45 nm for a surface velocity of 40 m/s and a load of 250 g, and the corresponding theoretical floating height based on the Burgdorfer equation was almost equal to (or slightly smaller than) the measured value [12]. On the other hand, the usual Reynolds' equation gave a floating height of 47 nm, about 4% larger than the experimental value. There is another report also that the Burgdorfer equation can be applied down to a floating height of 25 nm ($M = 8$) in a helium environment [16].

For even smaller film thicknesses of about 10–20 nm (i.e., $M \gg 1$), the Boltzmann equation, based on the kinetic theory of gases, is needed for the analyses of air films [15] [17] [18].

In such cases, the film thickness cannot be measured by usual optical interferometry because the film thickness is much smaller than the wavelength of light. Special methods are needed [21][25].

References

1. A.G.M. Michell, "Improvements in thrust and like bearings", British Patent No. 875 (1905).
2. A.G.M. Michell, "The Lubrication of Plane Surfaces", *Zeitschrift für Mathematik und Physik*, 52 (1905), Heft 2, pp. 123-137.
3. A. Kingsbury, "Thrust Bearings", US Patent No. 947242 (1910).
4. A.G.M. Michell, "Lubrication - Its Principles and Practice", *Blackie & Son Ltd.*, London and Glasgow, 1950.
5. A. Burgdorfer, "The Influence of the Molecular Mean Free Path on the Performance of Hydrodynamic Gas Lubricated Bearings", *Journal of Basic Engineering, Trans. ASME*, Vol. 81, March 1959, pp. 94-100.
6. W.A. Gross, "Gas Film Lubrication", *John Wiley & Sons, Inc., New York*, 1962.
7. H. Tahara, "Influence of the Pad-Deformation on the Michell Bearing Performance (First Report, Analysis of an Infinite Length, Spring Mounted Pad)" (in Japanese), *Trans. JSME*, Vol. 31, No. 231, November 1965, pp. 1731-1739.
8. H. Tahara, "Ditto (Second Report, Analysis of a Finite Length, Spring Mounted Pad)" (in Japanese), *Trans. JSME*, Vol. 31, No. 231, November 1965, pp. 1740-1749.
9. H. Tahara, "Ditto (Third Report, Experiments of the Centrally Supported Pads)" (in Japanese), *Trans. JSME*, Vol. 32, No. 234, February 1966, pp. 346-354.
10. H. Tahara, "Some Problems of Large Michell Thrust Bearings" (in Japanese), Journal of JSME, Vol. 69, No. 572, September 1966, pp. 1185-1194.
11. T. Yoshizawa, Y. Hori, H. Miura and M. Nemoto, "Studies on Air Floating Head Mechanism for Magnetic Storage Disks" (in Japanese), *Annual Report of the Engineering Research Institute, Faculty of Engineering, University of Tokyo*, Vol. 25, 1966, pp. 30-36.
12. Y. Mitsuya, "Molecular Mean Free Path Effects in Gas Lubricated Slider Bearings (Finite Element Solution)" (in Japanese), *Trans. JSME*, C, Vol. 44, No. 386, October 1978, pp. 3593-3602.
13. K. Ono, K. Kogure and Y. Mitsuya, "Dynamic Characteristics of Air-Lubricated Slider Bearings under Submicron Spacing" (in Japanese), *Trans. JSME*, C, Vol. 45, No. 391, March 1979, pp. 356-362.
14. Y. Mitsuya and R. Kaneko, "Molecular Mean Free Path Effects in Gas Lubricated Slider Bearings (Second Report, Experimental Studies)" (in Japanese), *Trans. JSME*, C, Vol. 46, No. 405, May 1980, pp. 542-549.
15. S. Fukui and R. Kaneko, "Analysis of Ultra-Thin Gas Film Lubrication based on Linearized Boltzmann Equation (First Report, Derivation of Generalized Lubrication

Equation)" (in Japanese), *Trans. JSME*, C, Vol. 53, No. 487, March 1987, pp. 829 - 838.
16. T. Ohkubo, J. Kishigami, S. Fukui and K. Yasuda, "Accurate Measurement of Gas Lubricated Slider Bearing Separation Using Visible Laser Interferometry" (in Japanese), *Trans. JSME*, C, Vol. 53, No. 487, March 1987, pp. 839 - 847.
17. S. Fukui and R. Kaneko, "Analysis of Ultra-Thin Gas Film Lubrication based on Linearized Boltzmann Equation (Second Report, Influence of Accommodation Coefficient)" (in Japanese), *Trans. JSME*, C, Vol. 53, No. 492, August 1987, pp. 1807 - 1814.
18. S. Fukui and R. Kaneko, "Analysis of Ultra-Thin Gas Film Lubrication Based on Linearized Boltzmann Equation: 1st Report - Derivation of a Generalized Lubrication Equation Including Thermal Creep Flow", *J. of Tribology, ASME*, Vol. 110, 1988, pp. 253 - 261.
19. N. Tagawa and M. Hashimoto, "Submicron Spacing of Self-Loading Flying Head Slider Mechanism (First Report, Fundamental Research)" (in Japanese), *Trans. JSME*, C, Vol. 55, No. 511, March 1989, pp. 774 - 779.
20. N. Tagawa and M. Hashimoto, "Ditto (Second Report, Experimental Research)" (in Japanese), *Trans. JSME*, C, Vol. 55, No. 518, October 1989, pp. 2625 - 2631.
21. H. Li, T. Kato and Y. Hori, "Accurate Measurement of the Flying Height of a Slider Taking into Account the Reflectance Variation of the Magnetic Disk Surface" (in Japanese), *Trans. JSME*, C, Vol. 55, No. 520, December 1989, pp. 3009 - 3013.
22. M. Tokuyama and S. Hirose, "Dyanamic Flying Characteristics of Magnetic Head Slider with Dust", *J. of Tribology, ASME*, Vol. 116, 1994, pp. 95 - 100.
23. D. Dowson, "History of Tribology (2nd Edition)", *Professional Engineering Publishing*, London and Bury St Edmunds, 1998, p. 657.
24. M. Ochiai and H. Hashimoto, "Static and Dynamic Characteristics of High Speed, Stepped Thrust Gas-Film Bearings (Theoretical Analysis Considering Fluid Inertia Forces)" (in Japanesse), *Trans. JSME*, C, Vol. 63, Vol. 613, September 1997, pp. 3249 - 3256.
25. Y. Mitsuya, K. Chihara and M. Yoshioka, "Novel Measurements of Flying Height and Attitudes Using Michelson Laser Interferometry", *IEEE Transaction on Magnetics*, Vol. 35 (1999), pp. 2338 - 2340.

5

Stability of a Rotating Shaft — Oil Whip

In high speed rotating machines such as turbines, blowers, and generators, vibration of a rotating shaft often hinders the smooth operation of the machine or even causes failure. Therefore, it is important to prevent such vibrations by investigating their nature. Among various kinds of vibrations in rotating machinery [32], oil whip, the subject of this chapter, is a self-excited vibration due to the oil film action of journal bearings and is one of the typical types of shaft vibrations. Since oil whip is closely connected with hydrodynamic lubrication, this chapter, dedicated to oil whip problems, was created.

Typical vibrations in rotating shafts, including oil whip, are listed below:

1. Resonance vibration at the critical speed
2. Parametric excitation due to the passage of balls or rollers of antifriction bearings [13]
3. Self-excited vibrations due to the internal damping of the shaft material [1]
4. Self-excited vibrations due to the oil film of journal bearings — oil whip
5. Flow-induced vibrations.

Item 1 is a forced vibration due to an imbalance of the shaft, and is a vibration to which attention must first be paid in most cases. When the rotating speed coincides with the critical speed of the shaft (the rotating speed corresponds to the natural frequency of the shaft), a violent vibration occurs as a result of resonance.

Item 2 is a vibration in a rotating shaft supported by rolling element bearings and is a kind of parametric resonance due to slight variation in bearing rigidity corresponding to the passing of balls or rollers in a loaded region.

Item 3 is a self-excited vibration due to the internal damping of the shaft material. It takes place over a wide range of shaft speed above the critical speed. Its frequency is equal to the natural frequency (critical speed) of the shaft regardless of its rotating speed. In the case of shafts made of materials of small internal damping such as steel, it may not be a problem, but in the case of assembled shafts, it will become a problem because the friction between the two parts in contact is equivalent to the internal damping of the material.

Item 4 is the subject of this chapter and is a self-excited whirling vibration due to oil film action in journal bearings. Since this may happen over a wide range of speeds above a certain threshold and may become a severe vibration, precautions must be taken. Its frequency is almost constant and is equal to the natural frequency of the shaft.

Item 5 is a self-excited whirling vibration due to fluid forces in the seal gap or blade tip gap of turbines and turbocompressors and its frequency is equal to the natural frequency of the shaft. Oil whip can be included in flow-induced vibrations in a wide sense.

5.1 Oil Whip

In experiments on a rotating shaft supported by journal bearings, Newkirk and Taylor (1925) found, under certain conditions, a new kind of severe vibration (or whirling) that was different from the vibration at the critical speed [2]. Since the vibration disappeared when the oil supply to the bearings was stopped and it resumed when the oil was supplied again, they concluded that the vibration was caused by the oil film in the bearing, and named it **oil whip**. Whip means a severe vibration of a shaft.

The phenomenon is described in more detail in Fig. 5.1.

When the rotating speed of a shaft is gradually increased, a large resonant vibration occurs in the shaft at the critical speed ω_1. The vibration diminishes, however, when the rotating speed passes the critical speed. Next, when twice the critical speed $2\omega_1$ is reached, a large vibration (or whirling) will appear as shown in Fig. 5.1a under certain conditions. When the shaft speed is increased further, this vibration will not diminish but it may continue as it was or it may become even larger, unlike resonant vibrations. This is typical of oil whip.

Oil whip was simply said to start at twice the critical speed in the early days. Later, however, some examples were reported in which the oil whip onset speed was somewhat or significantly higher than twice the critical speed as shown in Fig. 5.1b. This is often observed when the bearing pressure is high.

The features of oil whip are summarized in the following list:

1. When the rotating speed of a shaft is raised from zero, oil whip starts in many cases at twice the critical speed, as shown in Fig. 5.1a, and continues to exist beyond that speed. However, when the journal does not float up easily (for example when the bearing pressure is high or when the viscosity of the oil is low), oil whip may start at a speed higher than twice the critical speed, as shown in Fig. 5.1b. Even in this case, however, when the rotating speed is lowered, oil whip usually continues to exist down to twice the critical speed, i.e., the oil whip amplitude – shaft speed plot exhibits a **hysteresis loop** as shown in the figure. This phenomenon is sometimes called the inertia effect in oil whip.
2. The whirling speed (frequency) of a shaft in oil whip is almost constant, and is almost equal to the critical speed of the shaft.

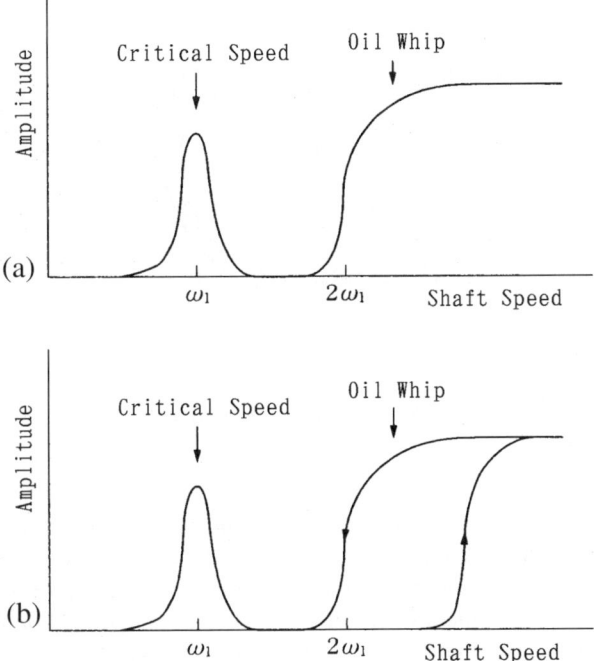

Fig. 5.1a,b. Oil whip. **a** typical oil whip, **b** oil whip with hysteresis

3. The whirling direction of a shaft in oil whip is the same as the rotating direction of the shaft.
4. Oil whip occurs easily when the journal floats up easily (i.e., the bearing pressure is low, or the viscosity of the oil is high).
5. Below twice the critical speed, the shaft may sometimes whirl quietly with little shaft bending. The whirling speed (frequency) of the shaft in this case is proportional to the rotating speed and is always equal to one-half of the rotation speed. Such whirling is called **oil whirl** or **half-speed whirl**.

Most of these features were found experimentally by Newkirk and Taylor and reported in their first paper [2], whereas the case shown in Fig. 5.1b was later reported by Newkirk and Lewis (1956)[9][10], Pinkus (1956) [11], and others.

Items 1 and 5 seem to indicate that twice the critical speed and half the rotating speed have special meanings in a journal bearing and that they are related to the nature of oil whip. The explanation for this by Newkirk and Taylor is as follows [2].

As shown in Fig. 5.2, the velocity of the oil in the oil film is equal to zero on the metal surface of the bearing and equal to the circumferential velocity of the journal on the journal surface. If a linear distribution of velocity between the two surfaces is assumed for simplicity, the average velocity of the oil in the oil film is equal to one-half the surface velocity of the journal, and is hence constant, at any section of

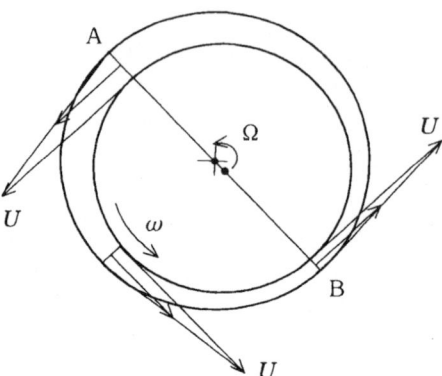

Fig. 5.2. Whirling speed of oil whip

the film. Therefore, the flow rate of oil at any arbitrary section of the oil film per unit width is proportional to the cross-sectional area. In other words, the volume of the oil that flows in through section A is much larger than the volume of the oil which flows out of section B and so, if incompressibility of the oil is assumed, the journal center must move in such a way that the excessive oil can be accommodated. As a result, the journal center circles around the bearing center. Now, since the difference of the volume of the oil which flows in through section A and that of the oil which flows out of section B per unit width must be equal to the volume swept by the journal in space, we have the following equation:

$$\frac{R_j \omega}{2}(c + e) - \frac{R_j \omega}{2}(c - e) = (2R_j)\, e\, \Omega$$

where R_j is the radius of the journal, $c = R_b - R_j$ is the radial clearance of the bearing, e is the eccentricity of the journal center, ω is the rotational speed of the journal, and Ω is the whirling speed of the journal.

The above equation gives the whirling speed of the journal center as follows:

$$\Omega = \frac{\omega}{2} \tag{5.1}$$

Thus, the journal will whirl in the same direction as that of rotation at an angular velocity of one-half the rotational speed. Therefore, if the rotational speed becomes twice the critical speed, the shaft will whirl just at the critical speed (natural frequency), and a large vibration due to resonance will take place. Taylor and Newkirk thought this was the mechanism of oil whip. Based on this idea, they referred to oil whip as oil resonance. As for the fact that large-amplitude resonance continues to exist even after the rotational speed is further increased, they explained that it is because the whirling speed does not increase any further because of the increase of friction due to the increase of amplitude.

Although their explanation describes the oil film action fairly well, there is a problem in explaining why a large vibration continues beyond twice the critical speed, because they regarded oil whip as a resonant phenomenon. Further, their theory cannot explain why in some cases oil whip starts only at a speed somewhat above twice the critical speed, and why in that case the amplitude change depicts a hysteresis loop. To explain these phenomena reasonably, it is necessary to treat oil whip as a self-excited vibration.

5.2 Oil Whip Theory

To explain oil whip theoretically as a self-excited vibration, it is necessary (1) to calculate the oil film force of the bearing, i.e., the force exerted by the oil film on the journal, then (2) to write down the equation of motion of the rotating shaft supported by the oil film force, and then (3) to examine the stability of the shaft (whether the shaft can rotate stably or becomes unstable, leading to a self-excited vibration) by applying a stability criteria to the equation of motion.

Robertson (1933) examined stability in this sense for the first time [3]. The oil film force he obtained under the assumption of an infinitely long bearing and Sommerfeld's boundary condition is as follows, being resolved into a component in the direction of eccentricity and that normal to it:

$$P_\kappa = -12\mu \left(\frac{R_j}{c}\right)^2 L R_j \frac{\pi \dot{\kappa}}{(1-\kappa)^{3/2}} \tag{5.2}$$

$$P_\theta = 12\mu \left(\frac{R_j}{c}\right)^2 L R_j \frac{\pi \kappa (\omega - 2\dot{\theta})}{(2+\kappa^2)\sqrt{1-\kappa^2}} \tag{5.3}$$

where μ is the coefficient of viscosity, R_j is the bearing radius, c is the radial clearance of the bearing, L is the bearing length, κ is the eccentricity ratio, θ is the attitude angle, and ω is the rotating speed of the shaft. The dots over κ and θ show time diffrentiation. He examined the shaft stability graphically using these expressions, and concluded that the speed limit of stability is always zero, meaning that the rotating shaft is always unstable, but this is not actually the case. Later, Poritsky(1953) [4] used the same oil film force and examined the stability mathematically and more rigorously and obtained the same result. No papers presented at that time could explain the oil whip phenomenon satisfactorily.

In calculating the oil film force, Hori (1955)[6, 7](1958)[14](1959)[15, 16] (1) used Gümbel's (or the half-Sommerfeld) boundary condition instead of Sommerfeld's boundary condition, which had been used until then, and, in examining the stability, (2) divided the shaft vibrations into a small vibration and a large vibration and obtained their stability limits separately, and (3) combined them to explain the process of the occurrence of oil whip shown in Fig. 5.1.

Later, more precise calculations under various conditions became possible with the development of computers. For example, Someya (1963)[17] (1964)[19] (1965) [23] performed many detailed numerical computations of this kind of problem in finite length bearings. He calculated not only the stability but the loci of journals and rotors. He also carried out calculations on rotors with imbalances.

Gotoda (1963) [18](1964) [20] also performed detailed numerical computations for the oil film characteristics and the stability in the case of finite length bearings. Funakawa and Tatara (1964) [22] calculated the oil film characteristics and stability using the short bearing approximation. Nakagawa and Aoki (1965) [24] obtained an approximate analytical solution for a finite length bearing under a certain assumption, and performed similar calculations.

Harada and Aoki (1971) [31] studied the stability of a shaft supported by turbulent journal bearings. Ono and Tamura (1968) [28] studied rotor stability using the boundary condition Section 3.1.4(e).

The major points of oil whip theories are explained here mainly following Hori's papers.

5.2.1 Oil Film Pressure

First, the dynamic Reynolds' equation, which takes the journal motion into consideration, is solved for the dynamic oil film pressure. This pressure is then integrated over the journal surface to obtain the dynamic oil film force that acts on the journal. In so doing, it is assumed that the journal motion is sufficiently slow, considering shaft vibrations near the stability limit.

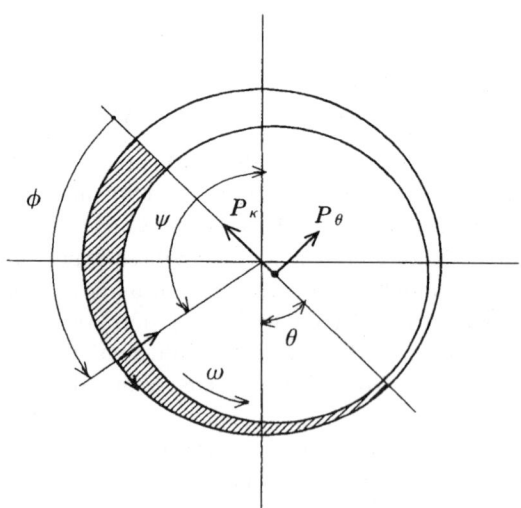

Fig. 5.3. Oil film and oil film force

If an infinitely long bearing is assumed in Fig. 5.3, the dynamic Reynolds' equation for the shaded part of the oil film can be written as follows:

5.2 Oil Whip Theory

$$\frac{1}{R_j^2}\frac{d}{d\psi}\left(h^3\frac{dp}{d\psi}\right) = 6\mu U \frac{1}{R_j}\frac{dh}{d\psi} + 12\mu\frac{\partial h}{\partial t} \quad (5.4)$$

where R_j is the radius of the journal and μ is the coefficient of vicosity. Further, h is the oil film thickness, U is the circumferential velocity of the journal, $\partial h/\partial t$ is time change of oil film thickness due to the journal motion, and these can be written as follows:

$$h = c\left(1 + \kappa\cos\phi\right), \quad \phi = \psi - \theta \quad (5.5)$$

$$U = R_j\,\omega \quad (5.6)$$

$$\frac{\partial h}{\partial t} = c\left(\frac{\partial \kappa}{\partial t}\cos\phi + \kappa\frac{\partial \theta}{\partial t}\sin\phi\right) \quad (5.7)$$

where c is the radial clearance, κ is the eccentricity ratio, and ω is the rotational speed of the shaft.

As the boundary condition for oil film pressure, Gümbel's condition is used, as stated before, assuming sufficiently slow motion of the journal, i.e., it is assumed that:

$$p = 0 \text{ at } \phi = 0 \text{ and } \phi = \pi \quad (5.8)$$

and, in the area where the calculated oil film pressure turns out to be negative, the pressure is replaced by zero.

Integration of Eq. 5.4 twice under the above conditions gives the pressure distribution $p(\phi)$ as follows [16]:

$$\frac{p(\phi)}{6\mu\omega(R_j/c)^2} = \left[J_2(\phi) - \frac{J_2(\pi)}{J_3(\pi)}J_3(\phi)\right]$$

$$+ \left[J_3^s(\phi) - \frac{J_3^s(\pi)}{J_3(\pi)}J_3(\phi)\right]\frac{2\dot{\kappa}}{\omega}$$

$$- \left[J_3^c(\phi) - \frac{J_3^c(\pi)}{J_3(\pi)}J_3(\phi)\right]\frac{2\kappa\dot{\theta}}{\omega} \quad (5.9)$$

where

$$J_k(\phi) = \int_0^\phi \frac{d\phi}{(1+\kappa\cos\phi)^k}, \quad J_k^c(\phi) = \int_0^\phi \frac{\cos\phi\,d\phi}{(1+\kappa\cos\phi)^k} \quad (5.10)$$

$$J_k^s(\phi) = \int_0^\phi \frac{\sin\phi\,d\phi}{(1+\kappa\cos\phi)^k} \quad (5.11)$$

These integrals can be calculated by using Sommerfeld's variable conversion as shown in Chapter 3, but here an alternative method by Okazaki shown below is used [5] [16].

The following integral is considered first. This can be found in handbooks of integral formulas:

$$J_{1\alpha}(\phi) = \int_0^\phi \frac{d\phi}{\alpha + \kappa\cos\phi} = \frac{2}{\sqrt{\alpha^2 - \kappa^2}}\tan^{-1}\left[\sqrt{\frac{\alpha-\kappa}{\alpha+\kappa}}\frac{\sin\phi}{1+\cos\phi}\right]$$

From the above formula, the following two integrals are easily obtained by differentiation under an integral sign with respect to α:

$$J_{2\alpha}(\phi) = \int_0^\phi \frac{d\phi}{(\alpha + \kappa \cos \phi)^2} = -\frac{\partial J_{1\alpha}}{\partial \alpha} = \frac{1}{\alpha^2 - \kappa^2}\left[\alpha J_{1\alpha} - \frac{\kappa \sin \phi}{\alpha + \kappa \cos \phi}\right]$$

$$J_{3\alpha}(\phi) = \int_0^\phi \frac{d\phi}{(\alpha + \kappa \cos \phi)^3} = -\frac{1}{2}\frac{\partial J_{2\alpha}}{\partial \alpha}$$
$$= \frac{1}{2(\alpha^2 - \kappa^2)^2}\left[(2\alpha^2 + \kappa^2)J_{1\alpha} - \frac{3\alpha\kappa \sin \phi}{\alpha + \kappa \cos \phi} - \frac{(\alpha^2 - \kappa^2)\kappa \sin \phi}{(\alpha + \kappa \cos \phi)^2}\right]$$

Letting $\alpha = 1$ in the above integrals yields the following three integrals $J_k(\phi)$:

$$J_1(\phi) = \int_0^\phi \frac{d\phi}{1 + \kappa \cos \phi} = \frac{2}{\sqrt{1-\kappa^2}} \tan^{-1}\left[\sqrt{\frac{1-\kappa}{1+\kappa}} \frac{\sin \phi}{1 + \cos \phi}\right] \quad (5.12)$$

$$J_2(\phi) = \int_0^\phi \frac{d\phi}{(1 + \kappa \cos \phi)^2} = \frac{1}{1-\kappa^2}\left[J_1(\phi) - \frac{\kappa \sin \phi}{1 + \kappa \cos \phi}\right] \quad (5.13)$$

$$J_3(\phi) = \int_0^\phi \frac{d\phi}{(\alpha + \kappa \cos \phi)^3} = \frac{1}{2(1-\kappa^2)^2}$$
$$\times \left[(2 + \kappa^2)J_1(\phi) - \frac{3\kappa \sin \phi}{1 + \kappa \cos \phi} - \frac{(1 - \kappa^2)\kappa \sin \phi}{(1 + \kappa \cos \phi)^2}\right] \quad (5.14)$$

From the above results, the following integrals $J_k^c(\phi)$ are easily obtained:

$$J_1^c(\phi) = \int_0^\phi \frac{\cos \phi}{1 + \kappa \cos \phi} d\phi = \frac{1}{\kappa}\left[\phi - J_1(\phi)\right] \quad (5.15)$$

$$J_2^c(\phi) = \int_0^\phi \frac{\cos \phi}{(1 + \kappa \cos \phi)^2} d\phi = \frac{1}{\kappa}\left[J_1(\phi) - J_2(\phi)\right] \quad (5.16)$$

$$J_3^c(\phi) = \int_0^\phi \frac{\cos \phi}{(1 + \kappa \cos \phi)^3} d\phi = \frac{1}{\kappa}\left[J_2(\phi) - J_3(\phi)\right] \quad (5.17)$$

It is easy to calculate the following integrals $J_k^s(\phi)$:

$$J_1^s(\phi) = \int_0^\phi \frac{\sin \phi}{1 + \kappa \cos \phi} d\phi = -\frac{1}{\kappa} \ln \frac{1 + \kappa \cos \phi}{1 + \kappa} \quad (5.18)$$

$$J_2^s(\phi) = \int_0^\phi \frac{\sin \phi}{(1 + \kappa \cos \phi)^2} d\phi = \frac{1}{1+\kappa}\frac{1 - \cos \phi}{1 + \kappa \cos \phi} \quad (5.19)$$

$$J_3^s(\phi) = \int_0^\phi \frac{\sin \phi}{(1 + \kappa \cos \phi)^3} d\phi = \frac{1}{2\kappa}\left[\frac{1}{(1 + \kappa \cos \phi)^2} - \frac{1}{(1 + \kappa)^2}\right] \quad (5.20)$$

The following constants can be easily obtained from the above results:

$$\frac{J_2(\pi)}{J_3(\pi)} = \frac{2(1 - \kappa^2)}{2 + \kappa^2}, \quad \frac{J_3^c(\pi)}{J_3(\pi)} = -\frac{3\kappa}{2 + \kappa^2}, \quad \frac{J_3^s(\pi)}{J_3(\pi)} = \frac{4(1 - \kappa^2)^{1/2}}{\pi(2 + \kappa^2)} \quad (5.21)$$

All the terms in Eq. 5.9 have thus been determined. Therefore, the pressure distribution is now available.

5.2.2 Oil Film Force

Multiplying the oil film pressure $p(\phi)$ of the preceding section by $\cos\phi$ and $\sin\phi$, and integrating them in the range of $\phi = 0 - \pi$ as shown below yield the dynamic oil film force under Gümbel's boundary condition. L is the bearing length and R_j is the journal radius:

$$P_\kappa = L R_j \int_0^\pi p(\phi) \cos\phi \, d\phi, \quad P_\theta = L R_j \int_0^\pi p(\phi) \sin\phi \, d\phi \qquad (5.22)$$

After some calculations that are much more complicated than those for Sommerfeld's boundary condition, the following results are obtained:

$$P_\kappa = -6\mu \left(\frac{R_j}{c}\right)^2 L R_j \left[\frac{2\kappa^2(\omega - 2\dot\theta)}{(2+\kappa^2)(1-\kappa^2)} + \frac{2\dot\kappa}{(1-\kappa^2)^{3/2}} \left\{ \frac{\pi}{2} - \frac{8}{\pi(2+\kappa^2)} \right\} \right] \qquad (5.23)$$

$$P_\theta = 6\mu \left(\frac{R_j}{c}\right)^2 L R_j \left[\frac{\pi\kappa(\omega - 2\dot\theta)}{(2+\kappa^2)(1-\kappa^2)^{1/2}} + \frac{4\kappa\dot\kappa}{(2+\kappa^2)(1-\kappa^2)} \right] \qquad (5.24)$$

where the dots over $\dot\theta$ and $\dot\kappa$ show time differentiation, and hence $\dot\theta$ is equal to the whirling speed of the shaft (Ω).

From a comparison of Eqs. 5.2 and 5.3 with Eqs. 5.23 and 5.24, the following can be seen. For $\dot\kappa = 0$, whereas Eq. 5.2 gives $P_\kappa = 0$, Eq. 5.23 generally gives a P_κ of finite value. The term $(\omega - 2\dot\theta)$ is not included in Eq. 5.2, but it is included in Eq. 5.23. Further, whereas $\dot\kappa$ is not included in Eq. 5.3, it is included in Eq. 5.24. These facts have important meanings in terms of bearing characteristics.

If the time derivatives are set to zero in Eqs. 5.23 and 5.24, they will become expressions for the oil film force in a stationary state, and are naturally the same as the expressions for the oil film force obtained under Gümbel's condition in Chapter 3. Therefore, the equilibrium position of the journal center is also the same as that obtained previously. The stability of small vibrations is to be considered in the neighborhood of this equilibrium point. Further, it is important that ω and $\dot\theta$ are in the expressions only in the combined form $(\omega - 2\dot\theta)$. This shows that no oil film force acts on the journal when it steadily whirls at a speed of half its rotating speed, and this is in agreement with the considerations of Newkirk and others in the previous section.

The oil film force in the case of the short bearing approximation can also be calculated by a similar method, and the results are given in the following form, similar to the above expressions [22]. These are useful because short bearings have been more and more frequently used in recent years:

$$P_\kappa = -\frac{1}{2}\mu \left(\frac{R_j}{c}\right)^2 \frac{L^3}{R_j} \left[\frac{2\kappa^2(\omega - 2\dot\theta)}{(1-\kappa^2)^2} + \frac{\pi\dot\kappa(1 + 2\kappa^2)}{(1-\kappa^2)^{5/2}} \right] \qquad (5.25)$$

$$P_\theta = \frac{1}{2}\mu \left(\frac{R_j}{c}\right)^2 \frac{L^3}{R_j} \left[\frac{\pi\kappa(\omega - 2\dot\theta)}{2(1-\kappa^2)^{3/2}} + \frac{4\kappa\dot\kappa}{(1-\kappa^2)^2} \right] \qquad (5.26)$$

The oil film force of a finite length bearing can also be written in the following form, similar to those of an infinitely long bearing and a short bearing [20]:

$$P_\kappa = -6\mu \left(\frac{R_j}{c}\right)^2 L R_j \left[(\omega - 2\dot\theta) P_\kappa^{(1)} + \dot\kappa P_\kappa^{(2)}\right] \qquad (5.27)$$

$$P_\theta = 6\mu \left(\frac{R_j}{c}\right)^2 L R_j \left[(\omega - 2\dot\theta) P_\theta^{(1)} + \dot\kappa P_\theta^{(2)}\right] \qquad (5.28)$$

In this case, $P_\kappa^{(1)}$, $P_\kappa^{(2)}$, $P_\theta^{(1)}$, and $P_\theta^{(2)}$ are functions of κ with the bearing dimensions as parameters. These are usually calculated numerically.

5.2.3 Linearization of the Oil Film Force

To discuss the linear stability of a shaft, the oil film force is linearized beforehand in the neighborhood of the equilibrium point of the journal center $O_{j0}(\kappa_0, \theta_0)$.

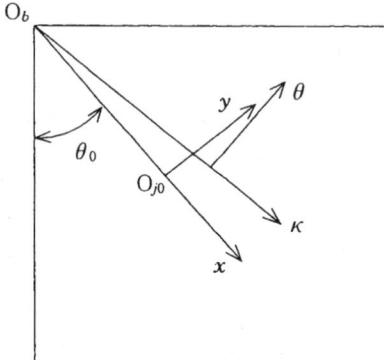

Fig. 5.4. Rectangular coordinates (radial direction, circumferential direction)

Equations 5.23 and 5.24 are used to express the oil film force. Then, variable substitutions $\kappa \Rightarrow \kappa_0 + \kappa$, $\theta \Rightarrow \theta_0 + \theta$ are made in these expressions, and assumptions that new variables κ, θ, and their time derivatives $\dot\kappa$ and $\dot\theta$ are so small that their second power, third power, ... and products can be disregarded give the linearized oil film force P_κ and P_θ as follows:

$$P_\kappa = -6\mu \left(\frac{R_j}{c}\right)^2 L R_j \times$$
$$\left[\frac{2\kappa_0^2 \omega}{(2+\kappa_0^2)(1-\kappa_0^2)} \left\{1 + \kappa\left(\frac{2}{\kappa_0} - \frac{2\kappa_0}{2+\kappa_0^2} + \frac{2\kappa_0}{1-\kappa_0^2}\right)\right\}\right.$$

5.2 Oil Whip Theory

$$-\frac{4\kappa_0^2\dot{\theta}}{(2+\kappa_0^2)(1-\kappa_0^2)} + \frac{2\dot{\kappa}}{(1-\kappa_0^2)^{3/2}}\left\{\frac{\pi}{2} - \frac{8}{\pi(2+\kappa_0^2)}\right\}\right] \quad (5.29)$$

$$P_\theta = +6\mu\left(\frac{R_j}{c}\right)^2 L R_j \times$$

$$\left[\frac{\pi\kappa_0\omega}{(2+\kappa_0^2)\sqrt{1-\kappa_0^2}}\left\{1 + \kappa\left(\frac{1}{\kappa_0} - \frac{2\kappa_0}{2+\kappa_0^2} + \frac{\kappa_0}{1-\kappa_0^2}\right)\right\}\right.$$

$$\left. - \frac{2\pi\kappa_0\dot{\theta}}{(2+\kappa_0^2)\sqrt{1-\kappa_0^2}} + \frac{4\kappa_0\dot{\kappa}}{(2+\kappa_0^2)(1-\kappa_0^2)}\right] \quad (5.30)$$

Now, to consider the journal motion in the rectangular coordinate system (x, y) shown in Fig. 5.4, let us transform the above components of the oil film force to the rectangular components P_x and P_y using the following expressions:

$$P_x = P_\kappa - P_{\theta 0}\frac{y_j}{c\kappa_0}, \quad P_y = P_\theta + P_{\kappa 0}\frac{y_j}{c\kappa_0} \quad (5.31)$$

where $P_{\theta 0}$ and $P_{\kappa 0}$ are the stationary values of the oil film force at the equilibrium point, i.e., the constant terms of Eqs. 5.29 and 5.30. Now, let us perform the variable transformation $\kappa \Rightarrow x_j/c$, $\theta \Rightarrow y_j/c\kappa_0$, $\dot{\kappa}_j \Rightarrow \dot{x}_j/c$, and $\dot{\theta}_j \Rightarrow \dot{y}_j/c\kappa_0$, and assume that $x_j, y_j, \dot{x}_j,$ and $\dot{y}_j$ are sufficiently small, considering small vibrations.

Then the oil film forces P_x and P_y can be written as follows:

$$P_x = P_{x0} + K_{xx}\frac{P_0}{c}x_j + K_{xy}\frac{P_0}{c}y_j + C_{xx}\frac{P_0}{\omega c}\dot{x}_j + C_{xy}\frac{P_0}{\omega c}\dot{y}_j \quad (5.32)$$

$$P_y = P_{y0} + K_{yx}\frac{P_0}{c}x_j + K_{yy}\frac{P_0}{c}y_j + C_{yx}\frac{P_0}{\omega c}\dot{x}_j + C_{yy}\frac{P_0}{\omega c}\dot{y}_j \quad (5.33)$$

where P_{x0} and P_{y0} in the above equations are the stationary values of the oil film force at the equilibrium point, and are given as follows:

$$P_{x0} = -6\mu\left(\frac{R_j}{c}\right)^2 R_j L \frac{2\kappa_0^2\omega}{(2+\kappa_0^2)(1-\kappa_0^2)} \quad (5.34)$$

$$P_{y0} = +6\mu\left(\frac{R_j}{c}\right)^2 R_j L \frac{\pi\kappa_0\omega}{(2+\kappa_0^2)\sqrt{1-\kappa_0^2}} \quad (5.35)$$

P_0 is their resultant $P_0 = \sqrt{P_{x0}^2 + P_{y0}^2}$ and is given as follows:

$$P_0 = 6\mu\left(\frac{R_j}{c}\right)^2 R_j L \omega \frac{\kappa_0\sqrt{\pi^2 - (\pi^2 - 4)\kappa_0^2}}{(2+\kappa_0^2)(1-\kappa_0^2)} \quad (5.36)$$

where c is the radial clearance of the bearing and ω is the angular velocity of the rotating shaft. The coefficients $K_{xx}, \cdots, C_{xx}, \cdots$ are nondimensional coefficients, and, as shown below, are given as functions of κ_0 only. This is important:

$$K_{xx} = -\frac{4\kappa_0}{\sqrt{\pi^2-(\pi^2-4)\kappa_0^2}}\left(\frac{1}{\kappa_0}-\frac{\kappa_0}{2+\kappa_0^2}+\frac{\kappa_0}{1-\kappa_0^2}\right) \quad (5.37)$$

$$K_{xy} = -\frac{\pi\sqrt{1-\kappa_0^2}}{\kappa_0\sqrt{\pi^2-(\pi^2-4)\kappa_0^2}} \quad (5.38)$$

$$C_{xx} = -\frac{2(2+\kappa_0^2)}{\kappa_0\sqrt{1-\kappa_0^2}\sqrt{\pi^2-(\pi^2-4)\kappa_0^2}}\left\{\frac{\pi}{2}-\frac{8}{\pi(2+\kappa_0^2)}\right\} \quad (5.39)$$

$$C_{xy} = +\frac{4}{\sqrt{\pi^2-(\pi^2-4)\kappa_0^2}} \quad (5.40)$$

$$K_{yx} = +\frac{\pi\sqrt{1-\kappa_0^2}}{\sqrt{\pi^2-(\pi^2-4)\kappa_0^2}}\left(\frac{1}{\kappa_0}-\frac{2\kappa_0}{2+\kappa_0^2}+\frac{\kappa_0}{1-\kappa_0^2}\right) \quad (5.41)$$

$$K_{yy} = -\frac{1}{2}C_{xy}, \qquad C_{yx} = C_{xy}, \qquad C_{yy} = 2-K_{xy} \quad (5.42)$$

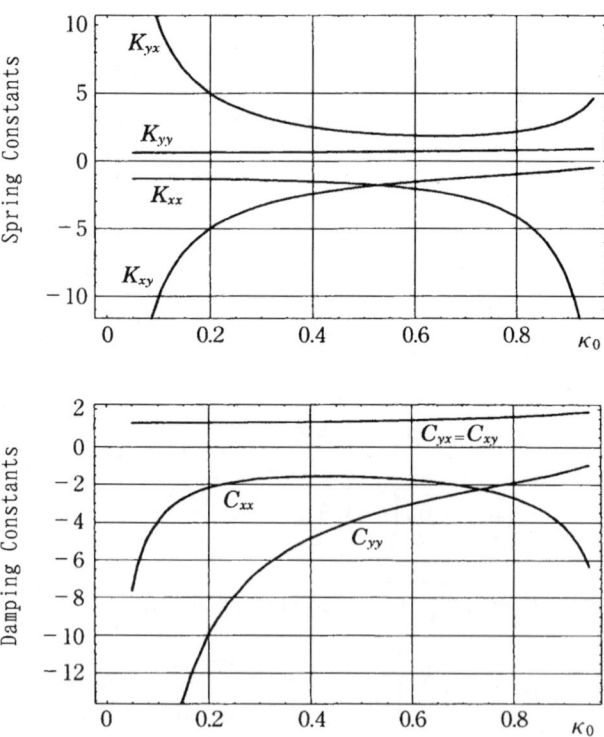

Fig. 5.5. Nondimensional spring constants and nondimensional damping constants of an oil film

Equations 5.32 and 5.33 can be written in an easier form to grasp as follows:

$$P_x = P_{x0} + k_{xx}x_j + k_{xy}y_j + c_{xx}\dot{x}_j + c_{xy}\dot{y}_j \tag{5.43}$$
$$P_y = P_{y0} + k_{yx}x_j + k_{yy}y_j + c_{yx}\dot{x}_j + c_{yy}\dot{y}_j \tag{5.44}$$

Or, in a matrix representation:

$$\begin{bmatrix} P_x \\ P_y \end{bmatrix} = \begin{bmatrix} P_{x0} \\ P_{y0} \end{bmatrix} + \begin{bmatrix} k_{xx} & k_{xy} \\ k_{yx} & k_{yy} \end{bmatrix} \begin{bmatrix} x_j \\ y_j \end{bmatrix} + \begin{bmatrix} c_{xx} & c_{xy} \\ c_{yx} & c_{yy} \end{bmatrix} \begin{bmatrix} \dot{x}_j \\ \dot{y}_j \end{bmatrix} \tag{5.45}$$

where

$$k_{ij} = K_{ij}\frac{P_0}{c}, \qquad c_{ij} = C_{ij}\frac{P_0}{\omega c} \tag{5.46}$$

In the above equations, k_{xx}, k_{xy}, k_{yx}, and k_{yy} are called **oil film spring constants** and c_{xx}, c_{xy}, c_{yx}, and c_{yy} are called **oil film damping constants**. These eight constants are collectively called **oil film constants**. Of these eight oil film constants, k_{xx}, k_{yy}, c_{xx}, and c_{yy} are called **diagonal terms** and k_{xy}, k_{yx}, c_{xy}, and c_{yx} are called **coupling terms**. The existence of a coupling term means that the direction of the force is different from that of the displacement and hence causes circumferential whirling of a shaft.

$K_{xx}, \cdots, C_{xx}, \cdots$ given by Eqs. 5.37–5.42 are called the **nondimensional spring constants** and the **nondimensional damping constants** of the oil film, respectively. Fig. 5.5 shows the nondimensional spring constants and the nondimensional damping constants graphically. The horizontal axes indicate the eccentricity ratio κ_0 of the equilibrium point.

In the case of the short bearing approximation and for a finite length bearing also, the oil film coefficients can be written in the form of Eq. 5.32 and Eq. 5.33, and it is known that the nondimensional coefficients $K_{xx}, \cdots, C_{xx}, \cdots$ are functions of κ_0 only. This is important.

It is recognized that the relation $C_{xy} = C_{yx}$ in Eq. 5.42 holds quite generally in various cases, in finite length bearings or under Reynolds' boundary condition, for example [54] [56].

5.2.4 Equations of Motion

By using the oil film force obtained in the preceding section, the equations of motion of a rotating shaft supported by journal bearings can be derived and dynamic characteristics of the rotating shaft can thereby be analyzed.

For simplicity, let us consider a system with one rotor and two bearings as shown in Fig. 5.6 (a system composed of a rotor supported by two journal bearings) and assume the following:

1. The shaft of the rotor is a thin bar of a circular section and its mass can be neglected.
2. A disk with mass is attached to the shaft at the center.

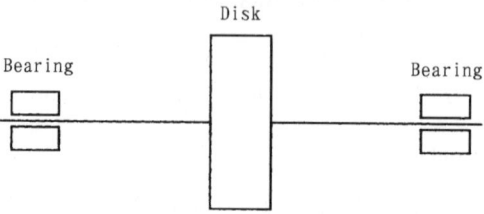

Fig. 5.6. A system with one rotor and two bearings

3. Both ends of the shaft are supported by journal bearings of the same specification.
4. The whole system is completely symmetrical with respect to the disk and has no imbalance.

The equations of motion of the system can be written as follows in the coordinate system of Fig. 5.7:

$$m\ddot{x} + k(x - x_j) = 0 \qquad (5.47)$$
$$m\ddot{y} + k(y - y_j) = 0 \qquad (5.48)$$
$$k(x - x_j) + P_x + P_1 \cos\theta_0 = 0 \qquad (5.49)$$
$$k(y - y_j) + P_y - P_1 \sin\theta_0 = 0 \qquad (5.50)$$

where m is the mass of the disk, k is the spring constant of the shaft, (x, y) and (x_j, y_j) are the coordinates of the disk center and the journal center, respectively; P_x and P_y are the x and y components of the oil film force P acting on the journal, respectively; and $P_1 = mg$ is the bearing load (g is the acceleration of gravity). From the balance of the bearing load and the oil film force at the equilibrium point, P_1 and P_0 of Eq. 5.36 must be equal, i.e., $P_1 = P_0$. Here, P and P_1 denote the sum of the oil film forces of the two bearings and the sum of the two bearing loads, respectively.

If the center of mass of the disk has a deviation δ, the right-hand side of Eqs. 5.47 and 5.48 should read $m\delta\omega^2 \cos\omega t$ and $-m\delta\omega^2 \sin\omega t$, respectively. ω is the angular velocity of the shaft.

5.2.5 Stability Limit

It is not easy to discuss the stability of the shaft in terms of the equation of motion of the previous section because the oil film force P_x and P_y are complicated nonlinear functions. Therefore, we divide the vibrations into two categories for which the equation of motion can be simplified, namely into sufficiently small vibrations and sufficiently large vibrations, and then discuss the stability of the two cases separately.

Small vibrations mean such vibrations of the journal center around the equilibrium point that the amplitude is sufficiently small compared with its eccentricity from

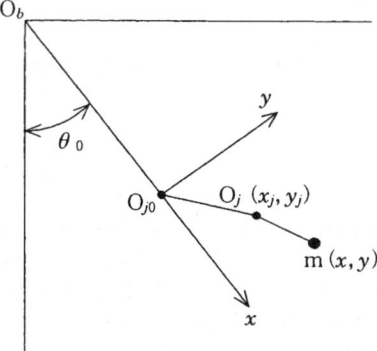

Fig. 5.7. Coordinates of the system (radial and circumferential direction)

the bearing center. The situation is shown in Fig. 5.8a,ba. In this case, the oil film force Eqs. 5.23 and 5.24 can be approximated by the linear expressions of Eqs. 5.32 and 5.33 in the neighborhood of the equilibrium point of the journal, as stated before.

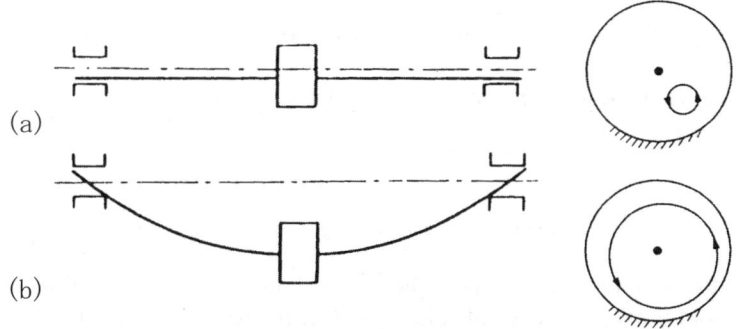

Fig. 5.8a,b. Small vibrations (**a**) and large vibrations (**b**) [14]

Large vibrations mean such vibrations (whirling) of the shaft that it bends considerably as shown in Fig. 5.8a,bb. In this case, the journal may tilt in the bearing, and the journal center inevitably circles around the bearing center for the majority of the bearing length (conical motion). In this case, the oil film force Eqs. 5.23 and 5.24 can be simplified by approximating the journal motion by a steady revolution. The stability limits (diverging criteria) of small vibrations and that of large vibrations are different.

By combining the stability limits of small vibrations and that of large vibrations, it is possible to explain the complicated process of the occurrence of oil whip.

a. Stability of Small Vibrations

Linearization of Equation of Motion   The following relations are obtained from Eqs. 5.47 and 5.48:

$$x_j = \frac{m}{k}\ddot{x} + x, \quad y_j = \frac{m}{k}\ddot{y} + y, \quad \dot{x}_j = \frac{m}{k}\dddot{x} + \dot{x}, \quad \dot{y}_j = \frac{m}{k}\dddot{y} + \dot{y} \quad (5.51)$$

By means of these relations, it is possible to eliminate the coordinates of the journal center (x_j, y_j) from the equation of motion Eqs. 5.47 and 5.48. Thus, substitution of Eqs. 5.49 and 5.50 into Eqs. 5.47 and 5.48, respectively, use of the oil film force for small vibrations Eqs. 5.32 and 5.33, and additional use of Eq. 5.51 give the following linearized form of Eqs. 5.47 and 5.48 on the coordinates of the disk center (x, y) only:

$$C_{xx}\frac{P_1}{k\omega c}\dddot{x} + \left(K_{xx}\frac{P_1}{kc} - 1\right)\ddot{x} + C_{xx}\frac{P_1}{m\omega c}\dot{x} + K_{xx}\frac{P_1}{mc}x$$
$$+ C_{xy}\frac{P_1}{k\omega c}\dddot{y} + K_{xy}\frac{P_1}{kc}\ddot{y} + C_{xy}\frac{P_1}{m\omega c}\dot{y} + K_{xy}\frac{P_1}{mc}y = 0 \quad (5.52)$$

$$C_{yx}\frac{P_1}{k\omega c}\dddot{x} + K_{yx}\frac{P_1}{kc}\ddot{x} + C_{yx}\frac{P_1}{m\omega c}\dot{x} + K_{yx}\frac{P_1}{mc}x$$
$$+ C_{yy}\frac{P_1}{k\omega c}\dddot{y} + \left(K_{yy}\frac{P_1}{kc} - 1\right)\ddot{y} + C_{yy}\frac{P_1}{m\omega c}\dot{y} + K_{yy}\frac{P_1}{mc}y = 0 \quad (5.53)$$

The stability of the rotor can be investigated by solving these equations simultaneously.

Stability Criterion   The above simultaneous equations can be written in the following general form:

$$A\dddot{x} + B\ddot{x} + C\dot{x} + Dx + E\dddot{y} + F\ddot{y} + G\dot{y} + Hy = 0 \quad (5.54)$$
$$a\dddot{x} + b\ddot{x} + c\dot{x} + dx + e\dddot{y} + f\ddot{y} + g\dot{y} + hy = 0 \quad (5.55)$$

When solutions of the form $x = \alpha e^{st}$ and $y = \beta e^{st}$ are assumed, the following equation must hold for the existence of solutions other than $x \equiv 0$ and $y \equiv 0$:

$$\begin{vmatrix} As^3 + Bs^2 + Cs + D, & Es^3 + Fs^2 + Gs + H \\ as^3 + bs^2 + cs + d, & es^3 + fs^2 + gs + h \end{vmatrix} = 0 \quad (5.56)$$

This is called the characteristic equation, and it will become the sixth-order equation below, if the determinant is developed:

$$A_0 s^6 + A_1 s^5 + A_2 s^4 + A_3 s^3 + A_4 s^2 + A_5 s + A_6 = 0 \quad (5.57)$$

where it is assumed that $A_0 > 0$. If $A_0 < 0$, then the sign of the whole equation will be changed so that $A_0 > 0$.

For the solutions x and y to be stable (i.e., they do not diverge), it is necessary and sufficient if the real part of all roots of the characteristic equation Eq. 5.57 are negative, and the **Routh–Hurwitz criterion** is known as a criterion for this. In the case of Eq. 5.57, it can be written as follows:

$$A_0, A_1, A_2, A_3, A_4, A_5, A_6 > 0 \tag{5.58}$$

$$\begin{vmatrix} A_1 & A_3 & A_5 & 0 & 0 \\ A_0 & A_2 & A_4 & A_6 & 0 \\ 0 & A_1 & A_3 & A_5 & 0 \\ 0 & A_0 & A_2 & A_4 & A_6 \\ 0 & 0 & A_1 & A_3 & A_5 \end{vmatrix} > 0 \tag{5.59}$$

$$\begin{vmatrix} A_1 & A_3 & A_5 \\ A_0 & A_2 & A_4 \\ 0 & A_1 & A_3 \end{vmatrix} > 0 \tag{5.60}$$

In other words, all the coefficients $A_0, A_1, \cdots$ must be positive and the two above determinants of these coefficients must also be positive.

In the case of Eqs. 5.52 and 5.53, the coefficients of the characteristic equation Eq. 5.57, $A_0, A_1, \cdots$ will be as follows:

$$A_0 = B_0 \frac{1}{\omega^2 \omega_1^4} \left(\frac{P_1}{mc}\right)^2$$

$$A_1 = B_1 \frac{1}{\omega \omega_1^4} \left(\frac{P_1}{mc}\right)^2 - B_2 \frac{1}{\omega \omega_1^2} \left(\frac{P_1}{mc}\right)$$

$$A_2 = 2B_0 \frac{1}{\omega^2 \omega_1^2} \left(\frac{P_1}{mc}\right)^2 + B_3 \frac{1}{\omega_1^4} \left(\frac{P_1}{mc}\right)^2 - B_4 \frac{1}{\omega_1^2} \left(\frac{P_1}{mc}\right) + 1$$

$$A_3 = 2B_1 \frac{1}{\omega \omega_1^2} \left(\frac{P_1}{mc}\right)^2 - B_2 \frac{1}{\omega} \left(\frac{P_1}{mc}\right)$$

$$A_4 = B_0 \frac{1}{\omega^2} \left(\frac{P_1}{mc}\right)^2 + 2B_3 \frac{1}{\omega_1^2} \left(\frac{P_1}{mc}\right)^2 - B_4 \left(\frac{P_1}{mc}\right)$$

$$A_5 = B_1 \frac{1}{\omega} \left(\frac{P_1}{mc}\right)^2, \qquad A_6 = B_3 \left(\frac{P_1}{mc}\right)^2$$

where $B_0, B_1, \cdots$ are as follows:

$$B_0 = C_{xx}C_{yy} - C_{xy}C_{yx}, \qquad \omega_1 = \sqrt{k/m}$$
$$B_1 = K_{xx}C_{yy} + K_{yy}C_{xx} - K_{xy}C_{yx} - K_{yx}C_{xy}$$
$$B_2 = C_{xx} + C_{yy}, \qquad B_3 = K_{xx}K_{yy} - K_{xy}K_{yx}$$
$$B_4 = K_{xx} + K_{yy}$$

When $K_{xx}, \cdots C_{xx}, \cdots$ are given by Eqs. 5.37–5.42, the actual calculations show that Eq. 5.58 of the Routh–Hurwitz criterion always holds, and if Eq. 5.59 holds, Eq. 5.60 always holds. Therefore, only Eq. 5.59 need be considered as a stability condition. If the determinant of Eq. 5.59 is developed, and the coefficients $A_0, A_1, \cdots$ are substituted into it, a comparatively simple result is obtained as follows. This is the stability criterion:

# 5 Stability of a Rotating Shaft — Oil Whip

$$\frac{1}{\omega^2}\left(\frac{P_1}{mc}\right) > K_1(\kappa_0)\left[K_2(\kappa_0) + \frac{1}{\omega_1^2}\left(\frac{P_1}{mc}\right)\right] \tag{5.61}$$

where $K_1(\kappa_0)$ and $K_2(\kappa_0)$ are given as follows:

$$K_1(\kappa_0) = \frac{B_1{}^2 - B_1 B_2 B_4 + B_2{}^2 B_3}{B_0 B_2{}^2}, \qquad K_2(\kappa_0) = \frac{B_2}{B_1} \tag{5.62}$$

Thus, finally, the shaft will be stable if Eq. 5.61 is satisfied.

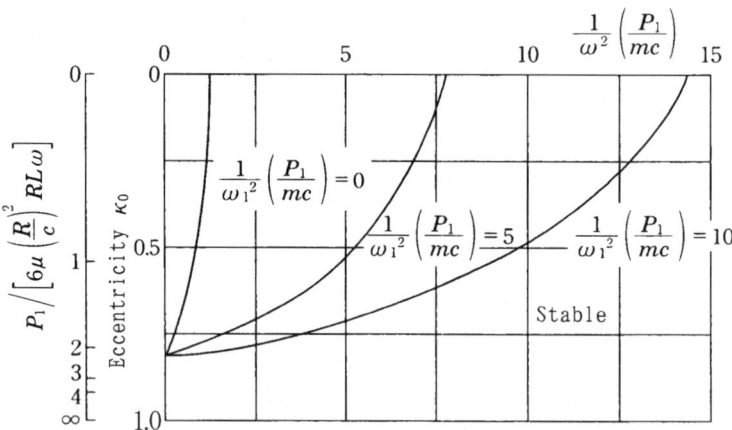

Fig. 5.9. Stability chart for an infinitely long bearing [14] [15]

Since K_1 and K_2 are functions of κ_0 only, Eq. 5.61 can be expressed in a chart as shown in Fig. 5.9. This is called a **stability chart**. The eccentricity ratio κ_0 is taken downward along a vertical axis, and on the other vertical axis to the left, a scale for the relation between nondimensional bearing load $P_1 / \left[6\mu(R/c)^2 RL\omega \right]$ and the eccentricity ratio κ_0 (cf. Eq. 5.36 where $P_0 = P_1$) is shown. The horizontal axis shows the nondimensional quantity $(1/\omega^2)(P_1/mc)$ which is made up of the rotational angular velocity, the bearing load, mass of the disk, and the bearing clearance. Three curves in the chart are the stability limit curves for three different values, 0, 5 and 10 of nondimensional parameter $(1/\omega_1{}^2)(P_1/mc)$, which is formed from the critical speed of the shaft, the bearing load, mass of the disk, and the bearing clearance. The leftmost curve corresponds to a rigid shaft.

To the lower right of each stability curve is the stable region and to the upper left is the unstable region. If the dimensions of the bearing and the shaft and the rotating speed, etc. are given, the position of the point corresponding to the operational condition is determined on the chart, and the stability can be judged by the side of the curve on which the point falls.

Although infinitely long bearings under Gümbel's boundary condition have been considered so far, short bearings or finite length bearings under other boundary con-

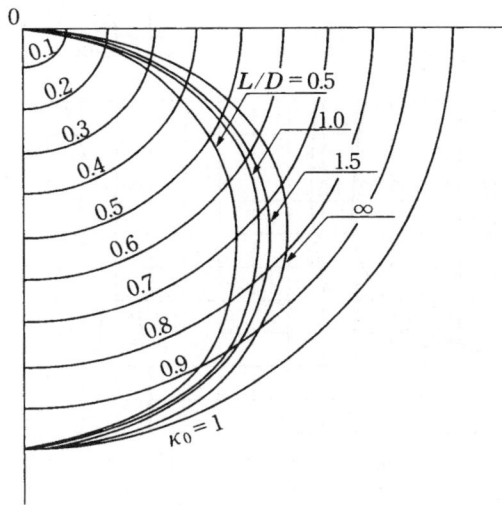

Fig. 5.10. Locus of the journal center for finite length bearings [18]

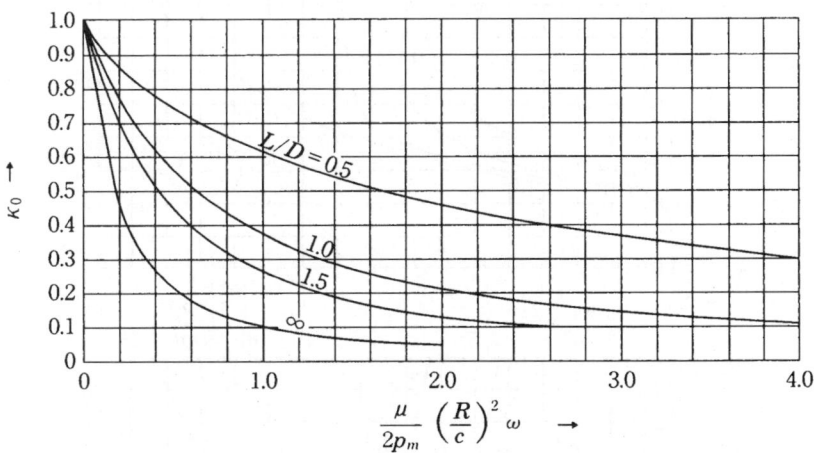

Fig. 5.11. Eccentricity of the journal for finite length bearings [18]

ditions can be discussed in a similar way. For example, when Gümbel's boundary condition is used for a finite length bearing, the locus of the journal center will be as shown in Fig. 5.10; the eccentricity ratio can be obtained from Fig. 5.11. The stability chart in this case is shown in Fig. 5.12 (Gotoda [18] [20]). As in the case of Fig. 5.9, the point corresponding to the dimensions of the bearings and rotating shafts, the rotating speed, and so forth is first determined on the stability chart, and then stability of the shaft is determined by the side of the stability curve on which the point falls.

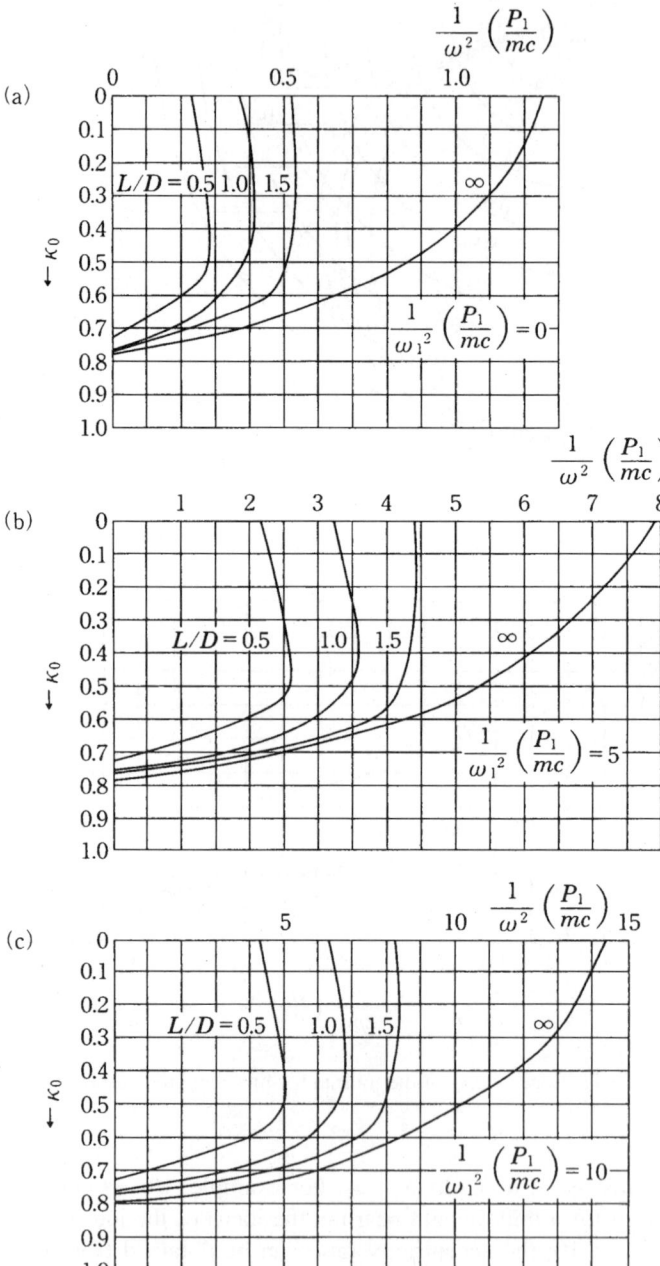

Fig. 5.12a-c. Stability charts for finite length bearings for three different values of the parameter $\frac{1}{\omega_1^2}\left(\frac{P_1}{mc}\right)$ [18]

From the stability chart, the following can be said:

1. The larger the eccentricity ratio, the more stable the shaft is. If the eccentricity ratio is larger than 0.8, in paticular, the shaft is always stable.
2. The higher the critical speed, the higher the stability limit is.
3. The smaller the (length L/diameter D) ratio of the bearings, the more stable the shaft is.
4. There is no such simple rule that the stability limit is equal to twice the critical speed.

It can be seen from the above analysis that a low oil viscosity, high bearing pressure, large bearing clearance, high critical shaft speed, and a small L/D ratio are recommended for high shaft stability.

b. Stability of Large Vibrations

When a shaft bends and whirls with a large amplitude, the journal center performs steady revolution around the bearing center for the major part of the bearing length, as shown in Fig. 5.8a,bb. Therefore, by setting the time derivative $\dot{\kappa}$ of the eccentricity ratio to be 0 in Eq. 5.24 of the oil film force, the circumferential component of the oil film force P_θ can be written in the following simple form:

$$P_\theta = K(\kappa) \cdot (\omega - 2\Omega) \tag{5.63}$$

where ω is the rotating speed of the shaft and $\Omega = \dot{\theta}$ is the whirling speed of the shaft. In the case of large vibrations (or whirling), stability means whether the whirling radius of the journal diverges or converges under the oil film force mentioned above, and in this case, twice the critical speed has an important meaning as seen from the above equation.

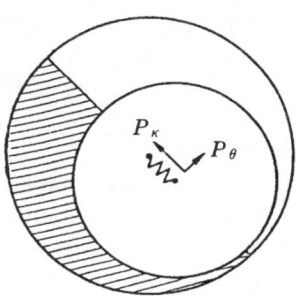

Fig. 5.13. Modeling of large vibrations [15]

For large vibrations, the shaft system can be modeled by a cylinder of mass m tied to the bearing center with a spring of spring constant k as shown in Fig. 5.13.

The equation of motion of the cylinder can be written as follows, the coordinates of the center of the cylinder being expressed by $Z = (c\kappa)e^{i\Omega t}$:

$$\ddot{Z} - K'(\kappa) \cdot (\omega - 2\Omega) \cdot iZ + \omega_1^2 Z = 0 \tag{5.64}$$

where $K' = K/m$ and $\omega_1 = \sqrt{k/m}$ (natural circular frequency of the cylinder, i.e., critical speed).

If $\Omega = \omega_1$ is assumed, or if it is assumed that the system is nearly in the state of the natural vibration, the equation has the following stationary solution when $\omega = 2\omega_1$:

$$Z = Ae^{i\omega_1 t}, \qquad \omega = 2\Omega = 2\omega_1 \tag{5.65}$$

This is because the damping term of Eq. 5.64 becomes zero in this case. When $\omega \neq 2\omega_1$, Z will diverge or converge, depending on whether the damping term is negative or positive, i.e., the stability limit of large vibrations (whirling) is given by:

$$\omega = 2\omega_1 \tag{5.66}$$

and the large vibrations will diverge or converge, depending on whether $\omega > 2\omega_1$ or $\omega < 2\omega_1$.

It should be noted, however, that divergence or convergence of whirling was discussed here under the assumption that the whirling of the journal around the bearing center already existed. If no whirling (vibrations) existed beforehand, large whirling does not necessarily occur even if the rotational speed reaches twice the critical speed.

5.2.6 Occurrence of Oil Whip — Hysteresis

As described in the previous section, the stability criteria for small vibrations and for large vibrations (whirling) are different. Combining these criteria provides a reasonable explanation of the hysteresis in the process of occurrence of oil whip.

Figure 5.14 shows a combination of one of the curves of Fig. 5.9, as an example, and the line of Eq. 5.66, i.e., the line for twice the critical speed. The latter is shown by a chained vertical line with the parameter ($2\omega_1$). Another chained line with the parameter (ω_1) shows the critical speed.

When the shaft speed increases from the stationary state, the shaft will be at the bottom of the bearing clearance ($\kappa = 1$) initially and finally floats up toward the bearing center. The point on Fig. 5.14 corresponding to the initial stationary condition is the lower extreme right (actually at infinity), and as the speed of rotation increases, the point moves toward the origin at the top left. The trajectory followed, however, is different depending on the conditions of the bearing as indicated by $\widehat{a_1 a_2}$, $\widehat{b_1 b_2}$ and $\widehat{c_1 c_2}$, which correspond to a light shaft, an intermediate shaft, and a heavy shaft, respectively.

The case of $\widehat{a_1 a_2}$ is considered first. While the operational point is around a_1, the shaft is still in the stable region. But beyond point A, the shaft enters the unstable

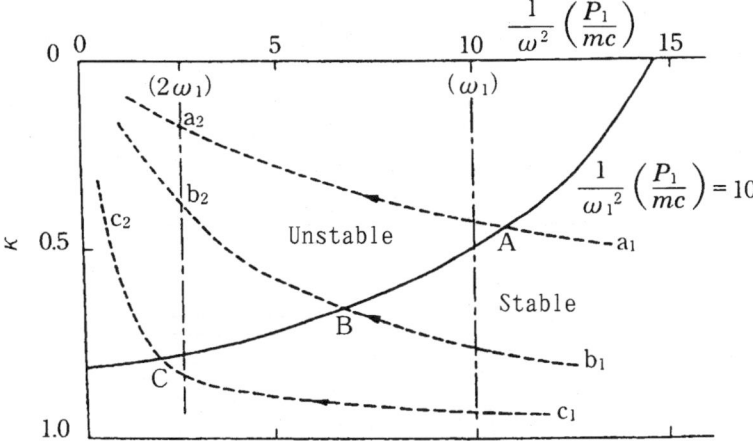

Fig. 5.14. Stability chart for small vibrations and for large vibrations [14]. $\widehat{a_1a_2}$ shows the path followed by a light shaft as the rotational speed increases, and $\widehat{b_1b_2}$ and $\widehat{c_1c_2}$ show the path for an intermediate and heavy shaft, respectively. Points A, B, and C indicate where these shafts enter the unstable zone

region and a half-speed whirl develops. However, since the line $(2\omega_1)$ has not yet been reached, the divergence condition of the whirling is not fulfilled and a large whirl does not develop. When $(2\omega_1)$ is reached, since the whirling speed of the half-speed whirl coincides with the natural frequency of the shaft, a large whirl occurs. Beyond this point, since the divergence condition for a large whirling is fulfilled, the whirling will diverge self-excitingly or at least continue. This is oil whip. The situation for a light shaft is shown in Fig. 5.15a-ca. The whirling speed of the shaft is equal to the critical speed (natural frequency), and the direction of whirling is the same as the direction of rotation of the shaft. At the critical speed en route, the resonance vibration and the half-speed whirl overlap each other.

Next, for a heavy shaft, as indicated by $\widehat{c_1c_2}$, even when twice the critical speed has been exceeded, the shaft is still in the stable region and even small vibrations do not occur. Therefore, although the divergence condition for whirling is fulfilled, the shaft remains stable. However, when point C is reached, the shaft becomes unstable and small vibrations will develop. Since the divergence condition for a whirl is already fulfilled at this time, it develops into oil whip immediately. When the rotational speed is further increased, the oil whip will continue to exist as in the case for a light shaft. Since oil whip, once established, continues to exist at speeds above twice the critical speed, when the rotating speed is lowered, oil whip will continue to occur down to twice the critical speed. Therefore, the routes of amplitude change during increasing and decreasing the shaft speed in this case are different, as shown in Fig. 5.15a-cc. This is the hysteresis phenomenon in oil whip dicussed at the begining of this chapter.

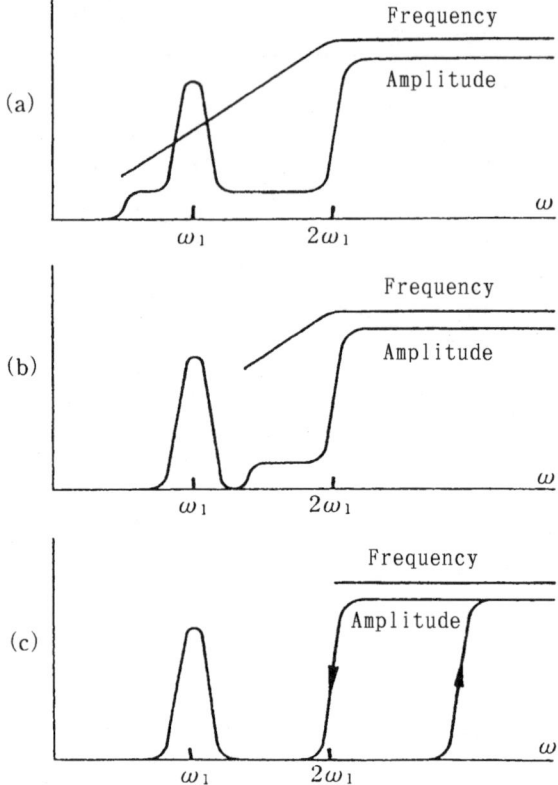

Fig. 5.15a-c. Occurrence of oil whip [14]. **a** in a light shaft (see $\widehat{a_1 a_2}$ in Fig. 5.14), **b** in an intermediate shaft, and **c** in a heavy shaft

$\widehat{b_1 b_2}$ is an intermediate case between the two above cases and the amplitude change will be as shown in Fig. 5.15a-cb.

Unlike the examples shown in Fig. 5.15a-c, there is an exceptional case in which the amplitude of oil whip decreases in the high speed region. This is probably due to the increase in whirling speed of the shaft with the increase in rotating speed of the shaft. This can be seen when the oil film force is extraordinarily large. The whirling speed in the typical case of, for example, $K/(2m\omega_1) = 0$, 1, and ∞ can be written as follows, respectively, where K is the coefficient in Eq. 5.63 [8].

$$\dot{\theta} = \omega_1 \quad \text{(constant, equal to natural frequency)} \quad (5.67)$$
$$\dot{\theta} = \sqrt{\omega_1/2}\sqrt{\omega} \quad \text{(proportional to square root of rotating speed)} \quad (5.68)$$
$$\dot{\theta} = \omega/2 \quad \text{(proportional to rotating speed)} \quad (5.69)$$

Equation 5.67 indicates typical self-excited vibrations (oil whip), Eq. 5.68 describes the above-mentioned case in which the amplitude falls gradually, and Eq.

5.69 describes the case of an extremely large oil film force that results in a kind of forced vibration rather than a self-excited vibration. In this case, the amplitude falls rapidly.

Oil whip has been so far considered on the basis of the primary critical speed ω_1. Concerning the secondary critical speed ω_2, "secondary oil-whip" is similarly possible at the speeds above twice the secondary critical speed. However, unless the primary oil whip has been attenuated beforehand, secondary oil whip will not be observed easily.

Large-Output Generators

Rotors of electric generators are usually operated at specific rated speeds, for example at 3000 rpm in the case of 50 Hz machines and 3600 rpm in the case of 60 Hz machines. The critical speeds of the rotors, however, differ from generator to generator depending largely on the generator output.

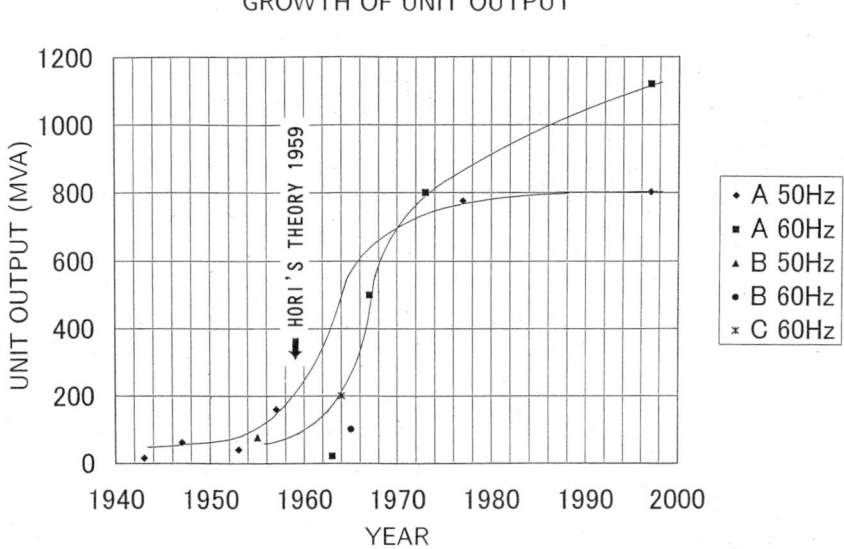

Fig. 5.16. Growth of unit output of generators [66]. A, B and C: manufacurers of generators

Generally speaking, the rotor of a generator of large output is longer than that for one of small output, the diameter being essentially unchanged. Increases in diameter would cause large centrifugal forces which could lead to destruction of the rotor. If the diameter is the same, a longer rotor has a lower critical speed. Therefore, the larger the output of the generator, the lower the critical speed of the rotor. If the critical speed becomes lower than 1500 rpm or 1800 rpm, i.e., lower than half the respective rated speed, the rotating speed of the shaft becomes higher than twice the critical speed and hence oil whip may occur. This will limit the maximum output of a generator. In fact, the maximum output of generators used to be kept below a certain level, say 100 MVA or at most 200 MVA, because of the possibility of oil whip.

However, it was shown by Hori [15](1959) for the first time that a rotor of any low critical speed can be operated at high speed without oil whip provided certain conditions (for example $\kappa_0 > 0.8$. cf. Fig. 5.14 and Fig. 5.15a-c) are satisfied by proper design of the bearings, e.g., by making the bearing of high enough pressure or equivalently by choosing a proper shape of bearing cross section (cf. Fig. 3.1a-f). Oil whip is no longer a barrier to the growth of output of generators.

The output of generators increased suddenly soon after the mechanism of oil whip was made clear, as shown in Fig. 5.16. This was a breakthrough in the design of large-output generators. Nowadays, the critical speeds of large generator rotors of the class of 1000 MW are often as low as 600 rpm. In other words, such generator rotors are operated at the speeds of five or six times the critical speed.

5.2.7 Coordinate Axes

The coordinate axes of Fig. 5.4 or Fig. 5.7 have been used up to now. On the other hand, the coordinate axes of Fig. 5.17 (horizontal and perpendicular axes) are widely used from the viewpoint of practical convenience. The components of the oil film force P_x and P_y in this case are obtained from the components P_x' and P_y' in the coordinate system of Fig. 5.7 by the following conversion:

$$P_x = P_x' \cos\theta_o - P_y' \sin\theta_o \tag{5.70}$$

$$P_y = P_x' \sin\theta_o + P_y' \cos\theta_o \tag{5.71}$$

Elastic coefficients and damping coefficients for the oil film can be obtained from these as before.

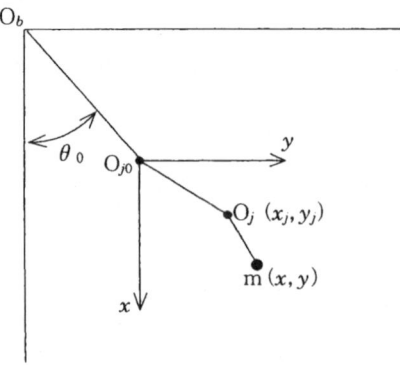

Fig. 5.17. Perpendicular and horizontal coordinates for a shaft system

A number of nondimensional elastic and damping coefficients of various types of bearings in the coordinate system of Fig. 5.17 as a function of the Sommerfeld number are given in bearing data handbooks [49]. Dynamic analyses of rotating shafts can be easily performed by using these data.

5.3 Stability of Multibearing Systems

The stability of the simplest systems with one rotor and two bearings has been considered so far. However, in the case of a large steam turbine generator, a generator is driven by, for example, four steam turbines. In such a case, several rotors connected by shaft couplings are supported by, sometimes, more than ten bearings. An example of such a multibearing system is shown in Fig. 5.18. In the case of multibearing systems, the **alignment** of bearings has a major influence on the stability of the system. This is because the load distribution to each bearing is a statically indeterminate problem in the case of multibearing systems, and if one of the bearings is displaced a little in the direction normal to the shaft (vertical or horizontal), the load distribution to each bearing changes a great deal and the stability of the shaft at low bearing loads is generally low. In the case of a system with one rotor and two bearings, displacement of a bearing hardly changes the load distribution, therefore stability does not change either.

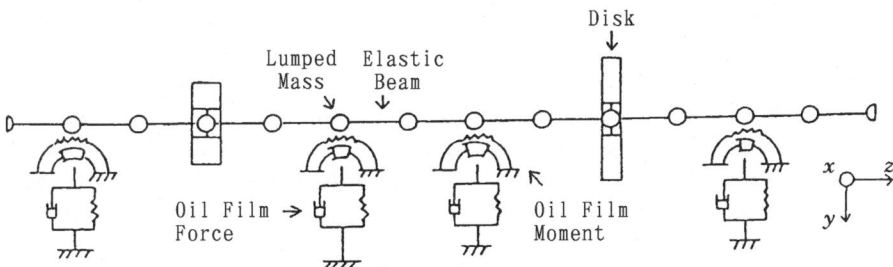

Fig. 5.18. Multibearing system [42]

Therefore, if alignment of the bearings at the time of installation is poor or alignment changes after installation as a result of distortion of the floor, for example, the stability of multibearing systems will often deteriorate. In such a case, a piece of thin metal is often inserted under a bearing to adjust the bearing height, and to improve the stability. This kind of adjustment of alignment has been usually performed according to the intuition of engineers in the field.

Holmes et al. [37] considered a two-rotor/four-bearing system made of two identical sets of one-rotor/two-bearing systems rigidly connected with a shaft coupling. They introduced a perpendicular displacement to one of the two bearings near the coupling and investigated how the stability of the system changed. One of their conclusions was that if the two rotors were of the same specifications, the stability of the system deteriorated as a result of displacement of the bearing, irrespective of whether it was upward or downward.

Kikuchi et al. [38] studied the influence of bearing displacement on shaft stability in the case of a shaft with a disk at the center supported by a bearing at the center and two others at the shaft ends.

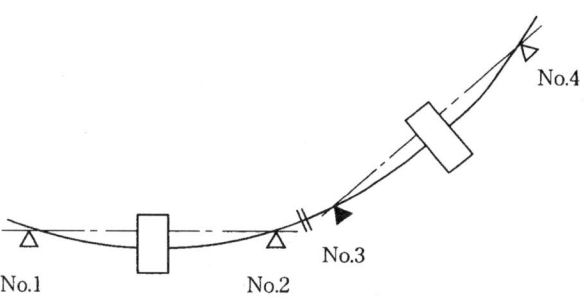

Fig. 5.19. A two-rotor/four-bearing system used for studying stability when bearing No.3 is displaced [42]

Hori et al. [39] and Nasuda et al. [42] considered a two-rotor/four-bearing system that consisted of two rigidly coupled sets of one-rotor/two-bearing system of the same specification and with different specifications, and studied the relation between stability and bearing displacement numerically. Figure 5.19 shows the system considered in which No.1 – No.4 are bearings. The positions of the four bearings were adjusted so that the positions (in the vertical and horizontal direction) and the inclinations of the shaft ends of the two rotors were equal at the position of the shaft coupling (catenary alignment), and this situation was taken as the standard alignment. Some displacement, in a vertical, a horizontal, or an inclined direction, was given to the No. 3 bearing, one of the bearings near the shaft coupling, and the relation between the bearing displacement, the displacement and its direction, and the change of the stability limit of the shaft was investigated.

For the calculations, the displacement given to the No.3 bearing was first assumed, then the load on each bearing and the eccentricity ratio of the journal was calculated by successive calculations (statically indeterminate problems). With the eccentricity ratios determined, the oil film characteristics (elastic coefficients, damping coefficients) could be calculated and then the linear calculations of stability were performed.

A transfer matrix method is used for vibration calculations of a shaft, so that a general shaft system can be dealt with, i.e., a given shaft is cut at suitable positions in the axial direction of the shaft into several shaft elements, and the state of shaft is expressed by eight quantities, i.e., displacement (u, v), inclination (φ, ψ), bending

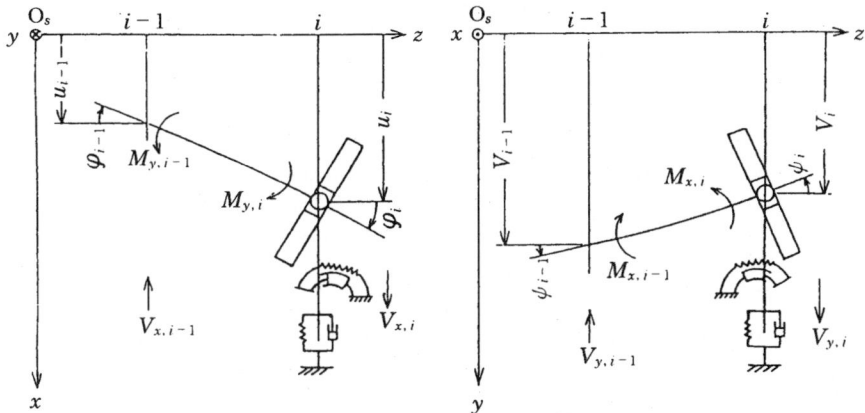

Fig. 5.20. Shaft elements for the transfer matrix method [42]

moment (M_x, M_y), and shear stress (V_x, V_y) at the cutting section. That is, the state at the ith section is expressed by the vector $Z_i = (u, \varphi, M_y, V_x, v, \psi, M_x, V_y)_i^T$. Then, the state vector at the ith and $i+1$th sections can be connected as follows by the element transfer matrix T_i which is determined by the strength-of-materials characteristics of the ith element and the oil film characteristics of the bearing:

$$Z_{i+1} = T_i \cdot Z_i \tag{5.72}$$

The shaft elements are shown in Fig. 5.20. This shows the most general case of a shaft element, and in actual cases either the rotor or the bearing may be missing. Applying the above equation to all the elements, we have the following relation between the state vectors of both ends of the shaft system, T being the product of all T_i:

$$Z_n = T \cdot Z_0 \quad \text{where} \quad T = \prod_{i=0}^{n} T_i \tag{5.73}$$

We assume that the solution of the equation is in the form $u = \bar{u} \cdot \exp(st)$, and that $V_x = V_y = M_x = M_y = 0$ are the boundary conditions at both ends of the shaft (both free ends). Then the following characteristic equation is derived from the above equation as the condition for all of u, v, θ_x, and θ_y are not equal to zero (nontrivial):

$$f(s) = 0 \tag{5.74}$$

The shaft will be stable if the real parts of all the roots of the characteristic equation are negative.

Figures 5.21a,ba and 5.21a,bb show two examples of the relation between the stability limit of a shaft thus obtained and the displacement of the bearing and direction of the displacement of the bearing. The rotors in Fig. 5.21a,ba consist of two

Table 5.1. Rotor dimensions

Rotor	l (mm)	m (kg)	I_p (kg·mm)	I_d (kg·mm)	N_c (rpm)	N_s (rpm)	L (mm)	R (mm)	c (mm)	μ (mPa·s)
A	1000	10	10	5	1893	3383	15	30	0.06	20
B	680	10	10	5	3376	5445	15	30	0.06	20

l, shaft length; m, disk mass; I_p, I_d, moment of inertia;
N_c, critical speed; N_s, stability limit; L, bearing length;
R, bearing radius; c, bearing clearance; μ, viscosity

identical A type rotors from Table 5.1 and the rotors in Fig. 5.21a,bb consist of a rotor B and a rotor A from the same table. The displacement of bearing No.3 normalized by the bearing clearance is taken on the horizontal axis and the symbol for each curve gives the direction of displacement of the bearing, taken every 30° from the vertical downward direction 0° to 150° in the direction of shaft rotation.

In the Fig. 5.21a,ba where two identical rotors are coupled, the stablity limit curve is symmetrical and the stability limit falls if the bearing is displaced from zero in any direction, especially in the 30° direction. In Fig. 5.21a,bb where two different rotors are coupled, the stability limit curve is asymmetrical and the stability limit is especially low in the direction of 30°. This requires special caution. The direction 30° may be related with the direction of the journal eccentricity. When the journal is displaced in the same direction but in the reversed sense, the stability limit goes up a little.

5.4 Influence of Earthquakes on Oil Whip

Although there has been much research on earthquake-resistant design of structures such as buildings, there is very little on that of machines, including rotating machines. However, machines such as generators are the cornerstones of human society, and their earthquake-resistivity is very important. The influence of an earthquake on oil whip in rotating machines is considered here.

As stated, so far the stability calculation of a rotating shaft has usually been performed using linear theories. However, this is inadequate to discuss the influence of a big disturbance such as an earthquake because of the nonlinearity of the oil film of a bearing. For example, a rotor in a large generator or steam turbine is operated below the stability limit but usually at a speed significantly higher than twice (typically five or six times) the critical speed. That is, the rotor is operated in the hysteresis region of oil whip. Although the rotor is stable in that region, it is stable only on the basis of a linear theory that considers small vibrations, and stability is not necessarily guaranteed under the influence of a large disturbance such as an earthquake. Since the condition for divergence of large amplitude whirl is already satisfied at speeds higher than twice the critical speed as stated before, a large disturbance such as an earthquake can trigger sudden violent oil whip in a previously stably running rotor. The influence of an earthquake in such a case has not been considered until now in

5.4 Influence of Earthquakes on Oil Whip

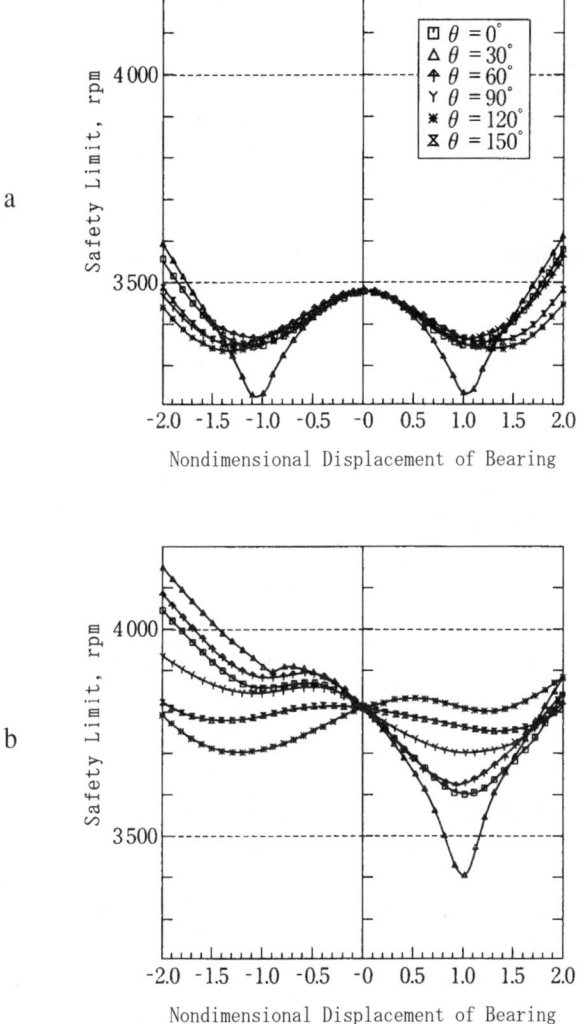

Fig. 5.21a,b. Stability of two-rotor/four-bearing systems. **a** with two A type rotors (see Table 5.1) and **b** with A and B type rotors, rotor A being on the No.3 and No.4 bearings [42]

the design of rotating machines such as generators. It is necessary to examine, at the design stage or even in operation, whether a rotor can remain in a stable state under the size of earthquake that can be expected.

Let us consider a rotor operated at a speed below the linear stability limit of oil whip but above twice the critical speed and examine its nonlinear stability by calculating the response of the rotor center to an earthquake using the Runge–Kutta–Gill method (Hori [51]).

5.4.1 Basic Equations

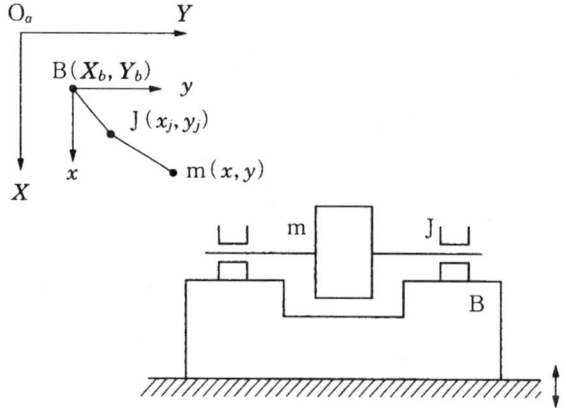

Fig. 5.22. Rotor on a pedestal with associated coordinates [51].

A symmetrical rotor supported by a bearing at either end of the shaft is installed on a pedestal as shown in Fig. 5.22. An earthquake occurs under the pedestal. With reference to the coordinate axes in the upper left part of Fig. 5.22, the equation of motion of the rotor can be written as follows (the x axis is taken downward):

$$m\ddot{x} + k(x - x_j) = -m\ddot{X}_B + mg \tag{5.75}$$
$$m\ddot{y} + k(y - y_j) = -m\ddot{Y}_B \tag{5.76}$$
$$k(x - x_j) + 2P_x = 0 \tag{5.77}$$
$$k(y - y_j) + 2P_y = 0 \tag{5.78}$$

where $\ddot{X}_B$ and $\ddot{Y}_B$ are the vertical and horizontal accelerations, respectively, coming from outside to the bearings through the pedestal and P_x and P_y are the oil film forces per bearing. If only small vibrations of a journal in the neighborhood of the equilibrium point are considered, the oil film force can be approximated by the linear formulae with the eight oil film coefficients. In this case, however, such linearizations are not allowed because the large vibrations caused by earthquakes are under consideration. Reynolds' equation must be solved for the position and the velocity of the journal at every short time step, then the oil film pressure obtained must be integrated over the journal surface at each time step to obtain the oil film force, and then the equation of motion must be solved at each time step repeatedly (Kato and Hori [52][53], Hori and Kato [55][57]).

5.4.2 Examples of Simulation

a. Response to a Sinusoidal Disturbance

Let us consider a rotor system with the following parameters:

$$L/D = 0.5, \quad B_p = 0.02, \quad \Omega_c = 1.0$$

where L/D is the slenderness ratio of the bearing, $B_p = (\mu L/\pi)(r/c)^3(g/W)\sqrt{c/g}$ is a bearing parameter, and $\Omega_c = \omega_c\sqrt{c/g}$ is an elasticity parameter of the rotor (where $\omega_c = \sqrt{k/m}$ is the critical speed). According to the linear theory, the stability limit of this system ω_{st} is as follows:

$$\omega_{st}/\omega_c = 4.714$$

For a rotating angular velocity of the rotor ω,

$$\omega/\omega_c = 3.2$$

is taken here because, as shown in Fig. 5.23a,ba, this is in the domain of hysteresis, the area of interest here. We are not interested in $\omega/\omega_c = 1.5$ in a linearly stable region or $\omega/\omega_c = 5.5$ in a linearly unstable region. Now, consider a rotor currently stably rotating at a speed of $\omega/\omega_c = 3.2$ (the journal center is at the equilibrium point) and suppose that one cycle of the following sinusoidal acceleration in the horizontal direction is suddenly applied to the pedestal:

$$\ddot{X}_B = 0 \tag{5.79}$$

$$\ddot{Y}_B = A\sin(\omega_e t), \quad 0 \leq t \leq 2\pi/\omega_e \tag{5.80}$$

$$A = 0.05G, \quad 0.1G, \quad 0.3G, \quad \omega_e = 0.8\omega_c \tag{5.81}$$

Three values of acceleration are considered (G = the gravitational acceleration). ω_e is the angular velocity of the sinusoidal wave of disturbance.

The response locus of the journal center calculated in the case of $A = 0.3G$ is shown in Fig. 5.23a,bb. It is seen that the shaft, which was running stably until the earthquake occurred, goes into an oil whip state, and the journal whirls to the limit of the clearance circle. For these calculations, the time step used by the Runge-Kutta-Gill method was $1/50 - 1/500$ of the natural frequency of the shaft. For $A = 0.05G$ and $A = 0.1G$, the disturbances were so small that they did not trigger oil whip.

5 Stability of a Rotating Shaft — Oil Whip

(a)

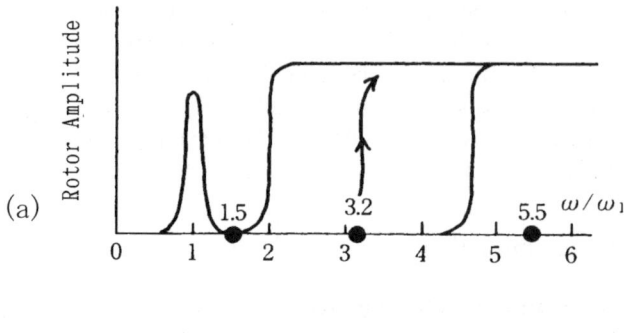

(b)

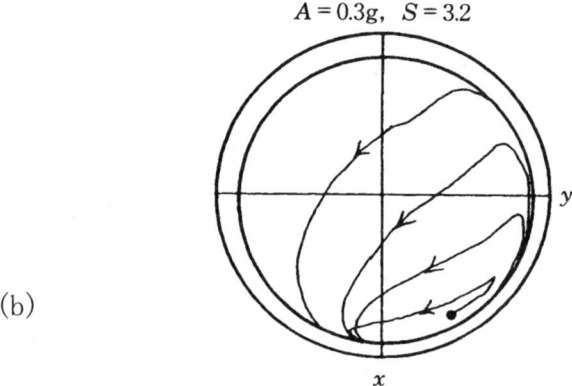

Fig. 5.23a,b. Response of rotor and journal to sinusoidal external forces [55]. **a** amplitude of oil whip as a function of rotational speed and **b** locus of the journal center

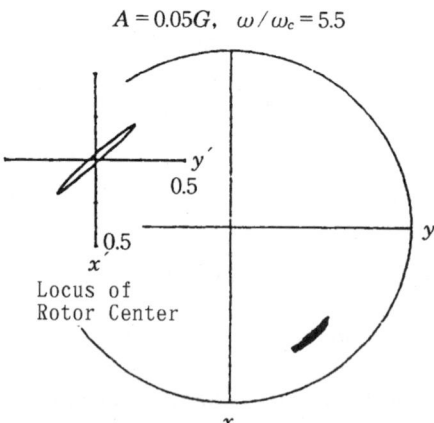

Fig. 5.24. Limit cycle in the unstable domain [55]

Thus, if a rotor that is rotating in the hysteresis domain is hit by a large external disturbance beyond a certain limit, the rotor will jump into an oil whip state, although nothing in particular happens if the disturbance is small. Since generator rotors are usually operated in the hysteresis domain, caution must be exercised.

An interesting phenomenon was noticed in the case of $\omega/\omega_c = 5.5$. At this speed, since the system is linearly unstable, a big oil whip should happen even if no disturbance exists. However, when a small disturbance ($A = 0.05G$) was applied, the response of the journal center rather unexpectedly settled into a small limit cycle as shown in the lower right quadrant of Fig. 5.24. This is probably due to the nonlinear nature of the oil film. The locus of the rotor center is shown in the upper left part of the figure.

b. Response to an Actual Seismic Wave

Let us consider a system in which the values of the system parameters are:

$$L/D = 0.5, \quad B_p = 0.01, \quad \Omega_c = (0.4)^{1/2}$$

The linear stability limit and the rated speed of rotation of this system are:

$$\omega_{st}/\omega_c = 8.490$$
$$\omega/\omega_c = 6.0$$

An actual example of a generator rotor with such parameter values is shown in Table 5.2.

Table 5.2. An example of generator rotors

Rotor	Bearing
$M = 44$ ton	$D = 662$ mm
$\omega_c = 600$ rpm	$L = 331$ mm
$\omega_{st} = 5094$ rpm	$c = 0.993$ mm
$\omega = 3600$ rpm	$\eta = 5.6$ mPa·s

As an example of an actual earthquake, the seismic wave in the east–west direction (maximum 175.9 Gal) and that in the vertical direction (maximum 102.9 Gal) of the Taft earthquake (1952, United States) is simultaneously applied to the system. Waves with frequencies of 10 Hz or less were prominent in this earthquake, which is a general tendency observed in many earthquakes. Figure 5.25 shows the results of repeated calculations of about 3500 steps (lasting 7 seconds), the time step being 1/50 of the natural frequency of the shaft (Hori et al. [55]). Figure 5.25a shows the locus of the journal center and Fig. 5.25b shows that of the rotor center (the scale of Fig. 5.25b is reduced). It turns out that a stable rotor in terms of linear calculations can go into the oil whip state because of an earthquake.

98   5 Stability of a Rotating Shaft — Oil Whip

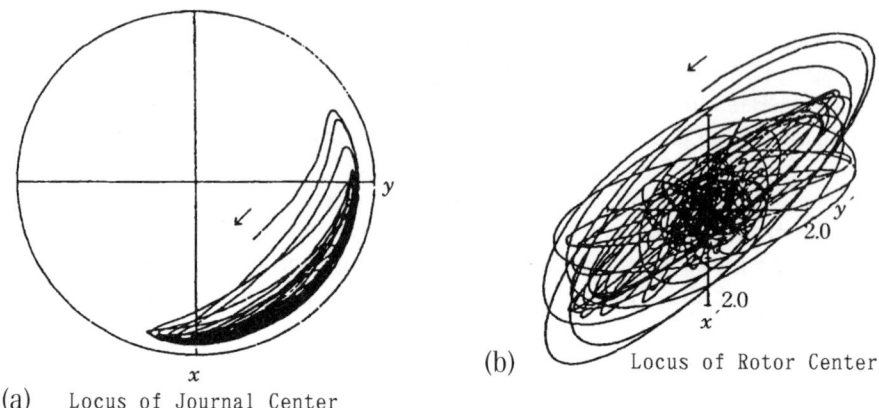

(a)   Locus of Journal Center   (b)   Locus of Rotor Center

Fig. 5.25a,b. Responses of journal and rotor to an external seismic force [55]. **a** locus of the journal center, **b** locus of the rotor center

Such a nonlinear approach is required for stability calculations that take the influence of an earthquake into consideration. Similar studies on the influence of an earthquake on oil whip were carried out for a multirotor system (Kato et al. [58] [60]) and multirotor systems with misalignment (Kato et al. [59] [61]).

5.5 Limit Cycle in an Unstable Domain

The tability of a rotating shaft supported by journal bearings is usually analyzed on the basis of linear theories, as stated. Namely, the oil film characteristics, which are nonlinear by nature, are linearized by assuming small vibrations, and the stability of a rotating shaft is analyzed by using the linearized oil film forces. On the other hand, it is known that a journal can rotate with an acceptably small limit cycle in the unstable domain as determined by the linear theories [43]. This phenomenon can be attributed to the nonlinear characteristics of the oil film [34] [41], and an example of such a limit cycle obtained by actual stepwise calculation of the journal locus was shown in Fig. 5.24 (Hori [51]). A big problem in such a calculation is, however, that an enormous amount of computing time is required to solve Reynolds' equation many times at every small step of journal movement.

5.5.1 Approximate Nonlinear Analysis of Journal Bearing Characteristics

A new approximate nonlinear solution method that includes up to secondary nonlinear terms developed by Malik et al. [47] is presented here with an example of the results. The computing time for this new method is much shorter than that in ordinary stepwise calculations.

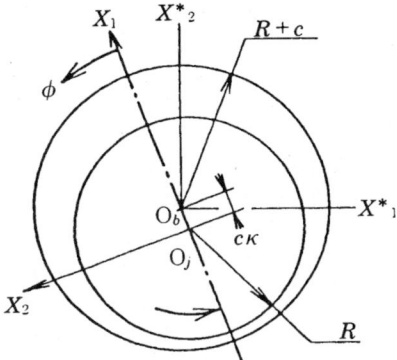

Fig. 5.26. Circular journal bearing [47]

Referring to Fig. 5.26, the film thickness equation is:

$$h = h_0 - X_1 \cos\phi - X_2 \sin\phi \tag{5.82}$$

where h_0 is the steady-state film thickness given by $h_0 = c(1 - \kappa\cos\phi)$ and X_1 and X_2 are the perturbation coordinates of the journal center relative to its steady state position O_j.

On the other hand, Reynolds' equation of the dynamically loaded bearing is:

$$\frac{\partial}{\partial\phi}\left(h^3\frac{\partial p}{\partial\phi}\right) + \frac{\partial}{\partial\zeta}\left(h^3\frac{\partial p}{\partial\zeta}\right) = 6\frac{\partial h}{\partial\phi} + 12\frac{\partial h}{\partial t} \tag{5.83}$$

On substituting Eq. 5.82 into Eq. 5.83, it becomes obvious that nonlinearity of the third order in X_1 and X_2 is inherently present in Eq. 5.83. However, the equation is linear in perturbed velocities. Thus, the pressure in a dynamically loaded bearing may conveniently be expressed, considering nonlinearity in X_1 and X_2 of second order only, as:

$$\begin{aligned}p =\ & p_0 + p_1 X_1 + p_2 X_2 + p_3 X_3 + p_4 X_4 \\ & + p_{11} X_1^2 + p_{12} X_1 X_2 + p_{22} X_2^2 + p_{13} X_1 X_3 \\ & + p_{14} X_1 X_4 + p_{23} X_2 X_3 + p_{24} X_2 X_4\end{aligned} \tag{5.84}$$

where

$$X_3 = \frac{dX_1}{dt}, \quad X_4 = \frac{dX_2}{dt} \tag{5.85}$$

In the above equation, p_0 is the steady state pressure, and $p_1, p_2, \cdots, p_{24}$ may be referred to as dynamic pressure coefficients. These coefficients are defined as:

$$p_1 = \left.\frac{\partial p}{\partial X_1}\right)_0, \quad p_2 = \left.\frac{\partial p}{\partial X_2}\right)_0, \quad \cdots \quad p_{24} = \left.\frac{\partial^2 p}{\partial X_2 X_4}\right)_0 \tag{5.86}$$

100   5 Stability of a Rotating Shaft — Oil Whip

where ")$_0$" implies that the values are determined at the equilibrium state.

With the above definitions, the governing equations of the pressure coefficients are obtained by successive differentiations of Eq. 5.83. As an example, the equation for p_{12} is:

$$\frac{\partial}{\partial \phi}\left(h_0^3 \frac{\partial p_{12}}{\partial \phi}\right) + \frac{\partial}{\partial \zeta}\left(h_0^3 \frac{\partial p_{12}}{\partial \zeta}\right) = -\frac{\partial}{\partial \phi}\left(h_{120}^3 \frac{\partial p_0}{\partial \phi}\right) - \frac{\partial}{\partial \zeta}\left(h_{120}^3 \frac{\partial p_0}{\partial \zeta}\right)$$
$$-\frac{\partial}{\partial \phi}\left(h_{10}^3 \frac{\partial p_2}{\partial \phi}\right) - \frac{\partial}{\partial \zeta}\left(h_{10}^3 \frac{\partial p_2}{\partial \zeta}\right) - \frac{\partial}{\partial \phi}\left(h_{20}^3 \frac{\partial p_1}{\partial \phi}\right) - \frac{\partial}{\partial \zeta}\left(h_{20}^3 \frac{\partial P_1}{\partial \zeta}\right) \quad (5.87)$$

and for p_{24} we have:

$$\frac{\partial}{\partial \phi}\left(h_0^3 \frac{\partial p_{24}}{\partial \phi}\right) + \frac{\partial}{\partial \zeta}\left(h_0^3 \frac{\partial p_{24}}{\partial \zeta}\right) = -\frac{\partial}{\partial \phi}\left(h_{20}^3 \frac{\partial p_4}{\partial \phi}\right) - \frac{\partial}{\partial \zeta}\left(h_{20}^3 \frac{\partial p_4}{\partial \phi}\right) \quad (5.88)$$

where

$$h_{120}^3 = \left(\frac{\partial^2 h^3}{\partial X_1 \partial X_2}\right)_0 = 6h_0 \cos \phi \sin \phi$$

$$h_{10}^3 = \left(\frac{\partial h^3}{\partial X_1}\right)_0 = -3h_0^2 \cos \phi \quad (5.89)$$

$$h_{20}^3 = \left(\frac{\partial h^3}{\partial X_2}\right)_0 = -3h_0^2 \sin \phi$$

Now consider the case of a balanced rigid rotor–shaft system, with a rotor of mass $2M$ at the midspan and the shaft supported at the ends on two identical journal bearings. The equations of motion of free translatory whirl of the shaft may be written as follows:

$$M\frac{dX_3}{dt} = P_1(X_1, X_2, X_3, X_4) - P_{10} \quad (5.90)$$

$$M\frac{dX_4}{dt} = P_2(X_1, X_2, X_3, X_4) - P_{20} \quad (5.91)$$

where P_1, P_2 and P_{10}, P_{20} are the oil film forces on the journal in the dynamic and steady state, respectively. i.e.,

$$P_1 = -\iint p \cos \phi \, d\phi d\zeta, \quad P_2 = -\iint p \sin \phi \, d\phi d\zeta \quad (5.92)$$

$$P_{10} = -\iint p_0 \cos \phi \, d\phi d\zeta, \quad P_{20} = -\iint p_0 \sin \phi \, d\phi d\zeta \quad (5.93)$$

Equations 5.90 and 5.91 are nonlinear in X_i (i = 1, 2, 3, 4); the components of oil film force P_1 and P_2 are implicitly nonlinear functions of perturbation coordinates and velocities.

In analogy with Eq. 5.84, the equations of motion Eqs. 5.90 and 5.91 may also be approximated, considering nonlinearities of second order in X_1 and X_2, as:

5.5 Limit Cycle in an Unstable Domain

$$M \frac{dX_3}{dt} = -d_{11}X_1 - d_{12}X_2 - d_{13}X_3 - d_{14}X_4$$
$$\quad - d_{111}X_1^2 - d_{112}X_1X_2 - d_{122}X_2^2 - d_{113}X_1X_3$$
$$\quad - d_{114}X_1X_4 - d_{123}X_2X_3 - d_{124}X_2X_4 \tag{5.94}$$

$$M \frac{dX_4}{dt} = -d_{21}X_1 - d_{22}X_2 - d_{23}X_3 - d_{24}X_4$$
$$\quad - d_{211}X_1^2 - d_{212}X_1X_2 - d_{222}X_2^2 - d_{213}X_1X_3$$
$$\quad - d_{214}X_1X_4 - d_{223}X_2X_3 - d_{224}X_2X_4 \tag{5.95}$$

where the d's are the dynamic coefficients, being defined as:

$$d_{i1} = -\left(\frac{\partial P_i}{\partial X_1}\right)_0, \quad d_{i2} = -\left(\frac{\partial P_i}{\partial X_2}\right)_0, \quad \ldots ,$$
$$d_{i24} = -\left(\frac{\partial P_i}{\partial X_2 \partial X_4}\right)_0 \quad i = 1, 2 \tag{5.96}$$

so that from Eqs. 5.84 and 5.92:

$$d_{1j} = \int \int p_j \cos \phi \, d\phi \, d\zeta \tag{5.97}$$

$$d_{2j} = \int \int p_j \sin \phi \, d\phi \, d\zeta \tag{5.98}$$

where j is a single or double subscript corresponding to the dynamic pressure coefficient.

In the usual linear analysis, the equations of motion are as follows, which are in agreement with Eqs. 5.94 and 5.95 if only the first-order terms are considered:

$$M \frac{dX_3}{dt} = -d_{11}X_1 - d_{12}X_2 - d_{13}X_3 - d_{14}X_4 \tag{5.99}$$

$$M \frac{dX_4}{dt} = -d_{21}X_1 - d_{22}X_2 - d_{23}X_3 - d_{24}X_4 \tag{5.100}$$

where $d_{11}, d_{12}, d_{21}, d_{22}$ are spring constants of the oil film and $d_{13}, d_{14}, d_{23}, d_{24}$ are damping coefficients.

5.5.2 Results of Analysis

A semianalytical finite element method is used for the solution of Reynolds' equation. Namely, the pressure distribution is expressed as a cos series in the axial direction, while one-dimensional isoparametric cubic elements are used in the circumferential direction. The accuracy of calculation of steady and dynamic characteristics is commensurate with the existing data [25] [26]. The solution of Eqs. 5.94 and 5.95 is carried out by the fourth-order Runge–Kutta method.

Table 5.3. Dynamic coefficients for the oil film of a journal bearing, d_{ij} ($j = 1-4$, linear terms; $j = 11-24$, nonlinear terms)

Aspect ratio $L/D = 1.0$, eccentricity ratio $\kappa = 0.6$, nondimensional bearing load $P_0 = 5.25$, linear stability limit $M_c = 36.51$

d_{ij}	$j = 1$	$= 2$	$= 3$	$= 4$
$i = 1$	3.4255	1.2681	5.9031	-2.1143
$= 2$	-2.4468	0.9712	-2.1143	2.5770

$= 11$	$= 12$	$= 22$	$= 13$	$= 14$	$= 23$	$= 24$
23.2595	-0.1478	-8.9769	-9.7937	1.9934	-3.0892	-2.4369
-7.7506	-0.8730	3.3898	2.0613	5.3423	-2.7628	3.0995

As an example, a circular journal bearing of aspect ratio $L/D = 1.0$, nondimensional bearing load $P_0 = 5.25$, and eccentricity ratio $\kappa = 0.6$ is considered. The stability limit obtained by the linear theory is, in terms of critical mass, $M_c = 36.51$. If M exceeds this value, the system will be unstable. The dynamic coefficients calculated in this case are shown in Table 5.3. Further, the approximate nonlinear transient responses of the journal as given by Eqs. 5.94 and 5.95 are shown in Fig. 5.27. Calculations were carried out in the three linearly unstable cases $M = 40.0$, $M = 41.5$, and $M = 42.5$; in all the cases $M > M_c$. The initial conditions for all three cases were:

$$X_1(0) = X_2(0) = 0, \quad X_3(0) = 0.01, \quad X_4(0) = 0 \quad (5.101)$$

It is interesting to see in the figure the existence of an asymptotically stable trajectory in the case of $M = 40.0$ and a limit cycle in the case of $M = 41.5$. In the case of $M = 42.5$, the response locus is diverging. The existence of a limit cycle in the unstable region is consistent with the predictions of nonlinear numerical analyses (cf. Fig. 5.24) and those of experiments.

If the case of a satisfactorily small limit cycle is regarded as stable, the above approximate nonlinear analysis gives a stability limit (M_c=41.5) about 14.7% higher than the stability limit of the linearized analysis (M_c=35.61). This fact indicates the difficulty of comparisons of theoretical and experimental stability limits.

The computing time required for the present approximate nonlinear analysis is roughly the same as that for linear analyses, and is about 1/100 of that for numerical nonlinear analyses. By using the approximate nonlinear analysis, therefore, the behavior of a journal near the stability limit, and thus the detailed structure of the stability limit, can be investigated.

5.6 Floating Bush Bearings

A floating bush bearing is a bearing that has a thin cylindrical bush floating freely between the fixed bush (bearing metal) and the journal as shown in Fig. 5.28. There

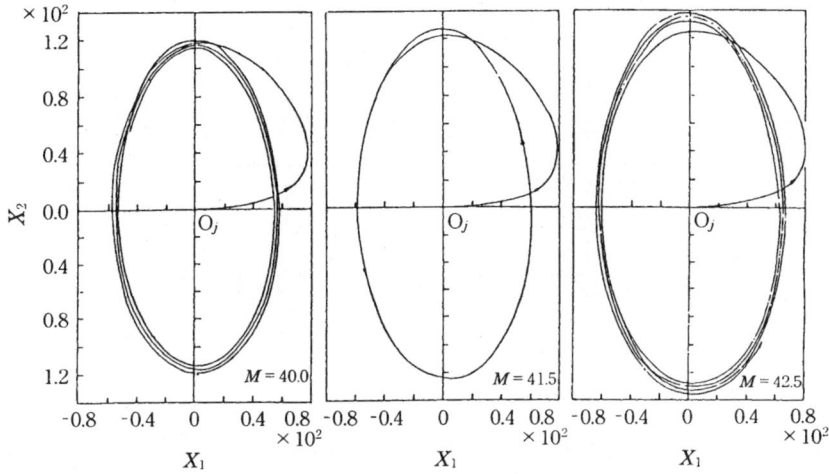

Fig. 5.27. Limit cycle in the unstable region [47]

are two oil films, inside and outside the free bush, which is called the floating bush. The floating bush bearing was first devised to reduce heat generation in a high speed journal bearing. In recent years, however, its vibration suppressing characteristics in high speed rotating shafts have attracted much attention.

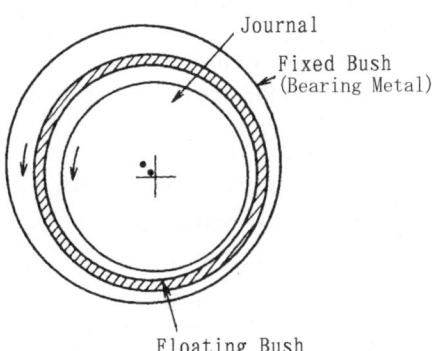

Fig. 5.28. Floating bush bearing

Oil whip suppression is one effect of floating bush bearings. In an experiment on a rotating shaft supported by floating bush bearings, it is reported that oil whip, which started at comparatively low speeds, was attenuated with the increase in rotational speed and finally disappeared (Tatara [30]).

Tanaka et al. [33] examined the effect of a floating bush bearing on oil whip.

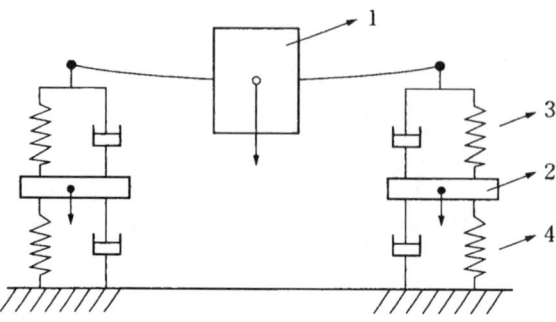

Fig. 5.29. Schematic of a rotor supported by floating bush bearings [33]. *1*, rotor; *2*, floating bush; *3*, inner oil film; *4*, outer oil film

A rotating shaft supported by floating bush bearings is schematically shown in Fig. 5.29. Numbers 1 and 2 in the figure are the rotor and the floating bush, respectively, and numbers 3 and 4 are the inside and outside oil films, respectively; an oil film is represented as a spring and a dash pot. The stability of the shaft system can be analyzed if the equation of motion of the system is formulated and Hurwitz's stability criterion is applied to its characteristic equation. In this case, however, the system is complicated and hence much calculation is required to get the characteristic equation, which is of the tenth order.

If Gümbel's condition is applied to both the inside and outside oil films in calculating the oil film force, it turns out that the following six nondimensional parameters are related to shaft stability:

ϕ = (weight of rotor)/(spring constant of shaft × inside clearance)
σ = (mass of floating bush)/(mass of rotor)
δ = (outer diameter of floating bush)/(inner diameter of the same)
β = (outside clearance)/(inside clearance)
v_1 = (shaft rotating speed)/$\sqrt{(\text{gravity acceleration})/(\text{inside clearance})}$
κ_1 = eccentricity ratio of journal in the inside clearance

An example of a stability chart with these parameters is shown in Fig. 5.30. In this case, $\phi = 0$, $\sigma = 0$, and $\delta = 1.32$; the horizontal axis of Fig. 5.30 being the bearing constant $\lambda_1 = \sqrt{g/c_1}(R_1/c_1)^2(\mu/2\pi p_m)(L/2R_1)^2$ and the vertical axis being the nondimensional rotating speed of the shaft $v_1 = \omega_1/\sqrt{g/c_1}$ (the subscript 1 refers to the inside oil film). To avoid a tedious calculation to determine the eccentricity ratio κ_1 of the journal in the inside clearance each time, κ_1 is eliminated and instead the bearing constant λ_1, which is directly calculable, is taken as the horizontal axis. The short bearing assumption is used here. Each stability limit curve is labeled with β = (outside clearance)/(inside clearance), $\beta = 0$ being an ordinary bearing without a floating bush. The lower side of each curve is the stable region. Although the stability limit curve of a floating bush bearing is quite complicated, as shown in Fig. 5.30, it can be said that, particularly in the domain of a small bearing constant (for example, when the bearing pressure is high), the stable region is significantly larger than that of ordinary journal bearings.

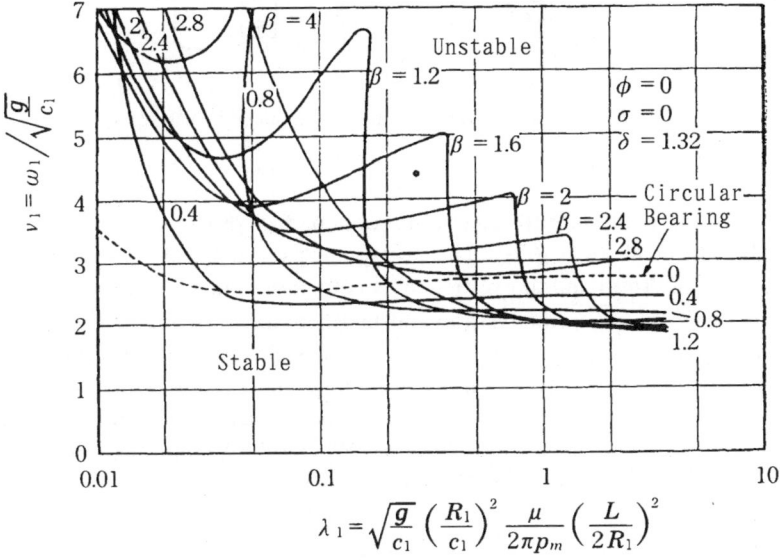

Fig. 5.30. Stability chart of floating bush bearing [33]

Thus the stability chart can explain how the onset speed of oil whip is raised by using floating bush bearings, but it cannot explain the above-mentioned phenomenon that the oil whip, once established, can disappear if the shaft speed becomes very high, for example, in the case of turbochargers.

An explanation for this phenomenon is given as follows.

In the case of a turbocharger, for example, the rotating speed of the journal is extremely high and the bearing pressure is low. Therefore it can be assumed that the inside oil film is in a concentric state. Further, because of the centrifugal force due to the extremely high rotating speed, the pressure at the journal surface is lower than that at the inner surface of the floating bush, and so a circular oil film rupture will occur at the end of the bearing (cf. Fig. 5.31) and proceed inward in the axial direction (Koeneke, Tanaka, et al. [62]). Then, since the driving torque on the floating bush decreases, the rotating speed ratio (floating bush rotating speed)/(journal rotating speed) of the floating bush will decrease. For the outside oil film, the ordinary Gümbel's boundary conditions is assumed.

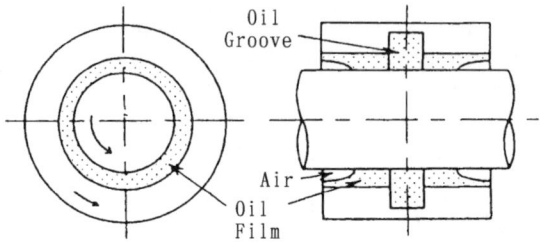

Fig. 5.31. Oil film in an extremely high speed bearing [64]

An example of a stability chart thus obtained is shown in Fig. 5.32 (Hatakenaka, Tanaka et al. [64] [65]). The horizontal axis is the bearing constant $\lambda = RL\mu(R/c)^2 \sqrt{g/c}/(mg)$ and the vertical axis is the nondimensional rotating speed of the journal $v_1 = \omega_1 \sqrt{c/g}$ (the bearing constant here differs from that of Fig. 5.30 by a constant). In the figure, the solid line shows the stability limit if oil film rupture in the inside film in the axial direction is considered, whereas the dashed line shows the stability limit when oil film rupture is not considered. The area between the thin and thick lines is the stable region for both cases. When oil film rupture of the inside film in the axial direction is considered (solid line), the stable region expands greatly in the high speed region (upward) when λ is around 10. If the rotating speed is raised when λ is small, the shaft is unstable at low speeds, then after passing a narrow stable zone, it becomes unstable again at higher speeds. When λ is large, even if the shaft is unstable at low speeds, it becomes stable over a wide range above a certain speed. This explains the above-mentioned phenomenon.

5.7 Three Circular Arc Bearings

A three circular arc bearing or a three arc bearing is known for its high stability. In the case of a vertical shaft, however, an additional condition α (offset factor) > 0.5 is necessary for shaft stability, as shown in the following.

In this section, the stator and the rotor of an electric motor for a geothermal water pump is discussed. Since the motor is installed deep underground in high pressure, high temperature water, the clearance between the stator and the rotor must be filled with oil to balance the pressure inside and outside the motor casing. Further, the shaft of the motor is vertical. Therefore, the shaft is very unstable and oil whip starts very easily. In such a case, it will be a good idea to use the principle of a three arc bearing for the inner surface of the stator, as shown in Fig. 5.33, to improve the stability (Hori et al. [48]).

5.7 Three Circular Arc Bearings   107

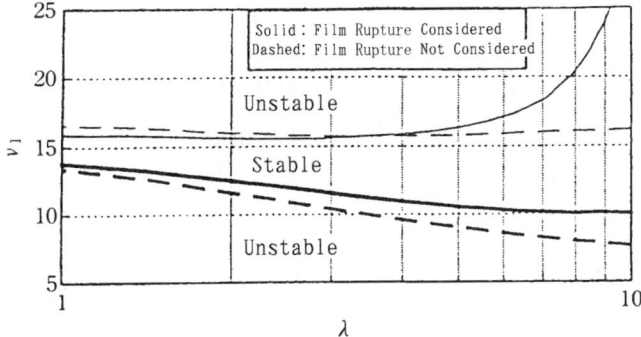

Fig. 5.32. Stability chart of an extremely high speed floating bush bearing [65]

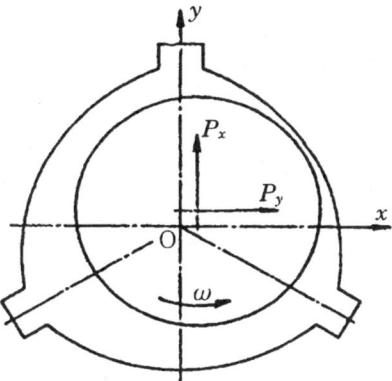

Fig. 5.33. Three circular arc bearing [48]

The geometrical features of a three circular arc bearing are identified by the following preload factor m_P and offset factor α:

$$m_P = 1 - c_b/(R_P - R) \tag{5.102}$$

$$\alpha = \beta/\chi \tag{5.103}$$

where c_b = minimum film thickness, R = rotor radius, R_P = radius of the arc of the stator, χ = angular extent of the arc, and β = angular extent of the converging region of the arc.

The way to proceed is to calculate the oil film force first, then to formulate the equation of motion of the rotor and apply Hurwitz's stability criterion to it as before. The stability limit of the rotor can then be obtained. Figure 5.34 shows a stability chart considering turbulent flow (cf. Chapter 8) because the clearance between the stator and the rotor is large. The vertical axis of Fig. 5.34 is the stability limit divided

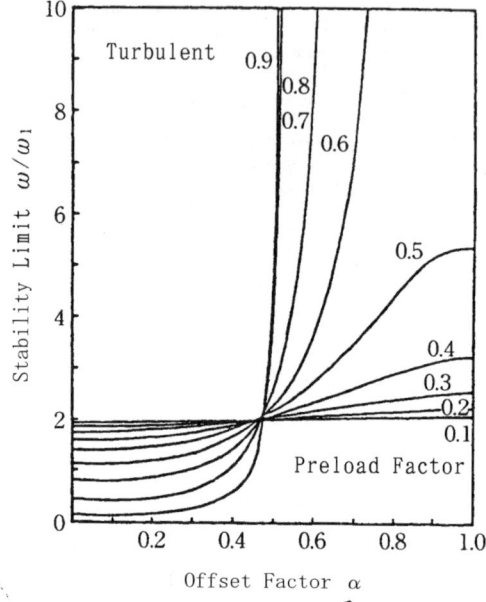

Fig. 5.34. Stability chart of a three arc bearing [48]

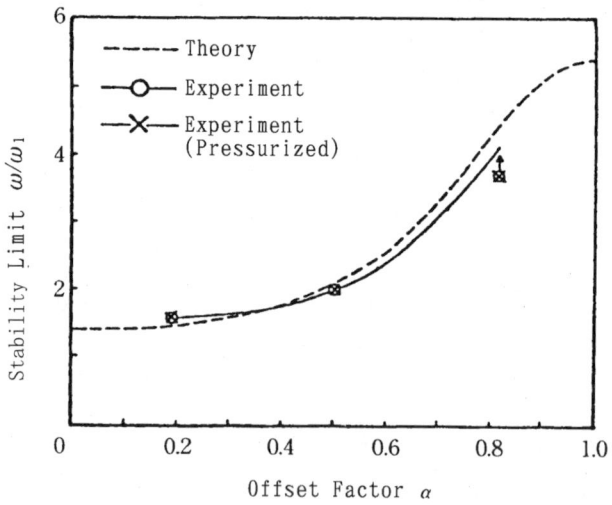

Fig. 5.35. Stability of a three arc bearing — comparison of theory and experiment [48]

by the critical speed and the horizontal axis is the offset factor. The figure shows that the stability limit is approximately twice the critical speed at any preload factor when the offset factor is 0.5. This means that, in this case, no benefit is expected from using a three arc cross section in the motor. If the offset factor is larger than 0.5, however, it can be seen that when the preload factor is large, the stability limit becomes fairly high. The stability of motors for geothermal water pumps can be improved by following this principle.

The above results of calculation coincide well with experiments, as shown in Fig. 5.35.

5.8 Porous Bearings

A porous journal bearing, in which the bush is made of an oil-soaked porous material, is widely used in light machines. The main purpose of using a porous bearing is to save the trouble of supplying oil and, instead, to perform lubrication by the oil that comes out of the bush with temperature rise. While porous bearings are usually used under boundary lubrication conditions, fluid lubrication may also be expected under certain conditions [45] and hydrodynamic analyses in such cases have actually been performed [12]. In this section, assuming a fluid lubrication condition, the stability of a rotating shaft in porous bearings is discussed (Hori and Okoshi [36]).

5.8.1 Governing Equations

The porous bearing shown in Fig. 5.36 is considered. A porous bush (porous bearing metal) is inserted in an impermeable housing and the lubricating oil is assumed to be incompressible.

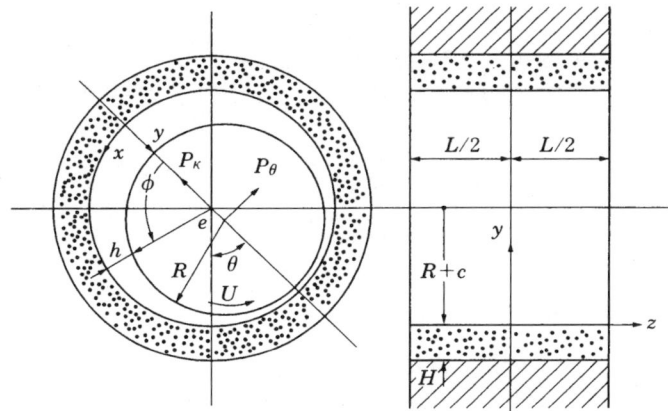

Fig. 5.36. Porous bearing [36]

Oil flow within the porous material of the bush is assumed to obey the following form of Darcy's law:

$$v = -\frac{\Phi}{\mu} \text{grad } p^* \tag{5.104}$$

where v is volume flow velocity vector per unit area, p^* is the pressure of oil in the porous material, μ is the coefficient of viscosity, and Φ is permeability.

Substituting Eq. 5.104 into the equation of continuity:

$$\frac{\partial v_x}{\partial x} + \frac{\partial v_y}{\partial y} + \frac{\partial v_z}{\partial z} = \nabla v = 0 \tag{5.105}$$

yields the following equation:

$$\frac{\partial^2 p^*}{\partial x^2} + \frac{\partial^2 p^*}{\partial y^2} + \frac{\partial^2 p^*}{\partial z^2} = \nabla^2 p^* = 0 \tag{5.106}$$

This is a Laplace equation with respect to the oil pressure p^* in the porous material.

To calculate the pressure in the lubricating film, the following Reynolds' equation with a correction term $(v_y)_{y=0}$ on the right-hand side is used:

$$\frac{\partial}{\partial x}\left(\frac{h^3}{12\mu}\frac{\partial p}{\partial x}\right) + \frac{\partial}{\partial z}\left(\frac{h^3}{12\mu}\frac{\partial p}{\partial z}\right) = \frac{U}{2}\frac{\partial h}{\partial x} + \frac{\partial h}{\partial t} - (v_y)_{y=0} \tag{5.107}$$

The correction term $(v_y)_{y=0}$ is the y component of the permeating velocity of the oil at the bush surface, and is given as follows from Eq. 5.104:

$$(v_y)_{y=0} = -\frac{\Phi}{\mu}\left(\frac{\partial p^*}{\partial y}\right)_{y=0} \tag{5.108}$$

Then, if a short bearing is assumed for simplicity, the basic equations for a porous bearing are as follows:

$$\frac{\partial}{\partial z}\left(\frac{h^3}{12\mu}\frac{\partial p}{\partial z}\right) = \frac{U}{2}\frac{\partial h}{\partial x} + \frac{\partial h}{\partial t} + \frac{\Phi}{\mu}\left(\frac{\partial p^*}{\partial y}\right)_{y=0} \quad \text{(oil film)} \tag{5.109}$$

$$\frac{\partial^2 p^*}{\partial y^2} + \frac{\partial^2 p^*}{\partial z^2} = 0 \quad \text{(porous bush)} \tag{5.110}$$

where $p = p^*$ must hold at the bush surface ($y = 0$).

5.8.2 Stability of a Shaft System

Under some appropriate assumptions on the pressure gradient in the porous bush [12], Eqs. 5.109 and 5.110 are solved simultaneously and loci of the journal center are obtained as shown in Fig. 5.37, the parameter Φ being permeability.

$\Phi = 0$ indicates the case of an impermeable bush, and the journal locus coincides with the ordinary journal locus in a short bearing. For $\Phi > 0$, the bush is permeable

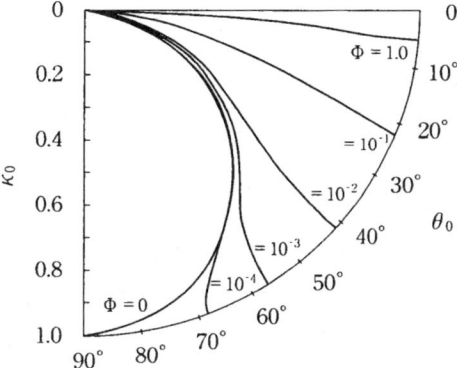

Fig. 5.37. Locus of the journal center in a porous bearing [36]

and since the lubricating oil soaks into the bush as a result of pressure in these cases, the oil film thickness decreases. If the bearing load is fairly high and the shaft speed is fairly low, the journal will contact the bush surface (the eccentricity ratio becomes 1). In this case, unlike the case of $\Phi = 0$, the journal locus intersects the bush surface at an angle.

The stability charts for rotating shafts supported by porous bearings in the same way as in Fig. 5.6 are shown in Fig. 5.38. The curves in the charts show the stability limit, the right side of each curve being the stable region and left side the unstable region. The parameter for each curve in the figure is $(1/\omega_1^2)(P_1/mc)$, which is the same as that in the previous stability charts.

Figure 5.38c shows the stability in the case $\Phi = 0$, i.e., an impermeable bush, and is identical to that of usual short bearings [22]. Figure 5.38b shows the stability in the case of $\Phi = 10^{-4}$, i.e. for some permeability, and 5.38a shows the case for $\Phi = 10^{-2}$, i.e., still larger permeability. The shapes of the stability limit curves in these charts differ greatly because of the difference in permeability. It is interesting to compare these three cases. Although the position of the stability limit curves are not very different for small eccentricity ratios ($\kappa < 0.6$), for large eccentricity ratios ($\kappa > 0.6$), both the position and the shape of the stability limit curves are very different. In Fig. 5.38a, particularly, where the permeability Φ is large, the stable region is very small. This is probably because the damping effect of the oil film decreases when the eccentricity ratio is high, since lubricating oil soaks into the bush because of the high pressure generated. It is interesting to see in Fig. 5.38b the process of transition of the stability curve from that in 5.38c to that in 5.38a.

5.9 Chaos in Rotor–Bearing Systems

In Section 5.4, it was shown by calculation that a shaft rotating stably in the hysteresis domain of oil whip can jump into the oil whip state as a result of a large disturbance,

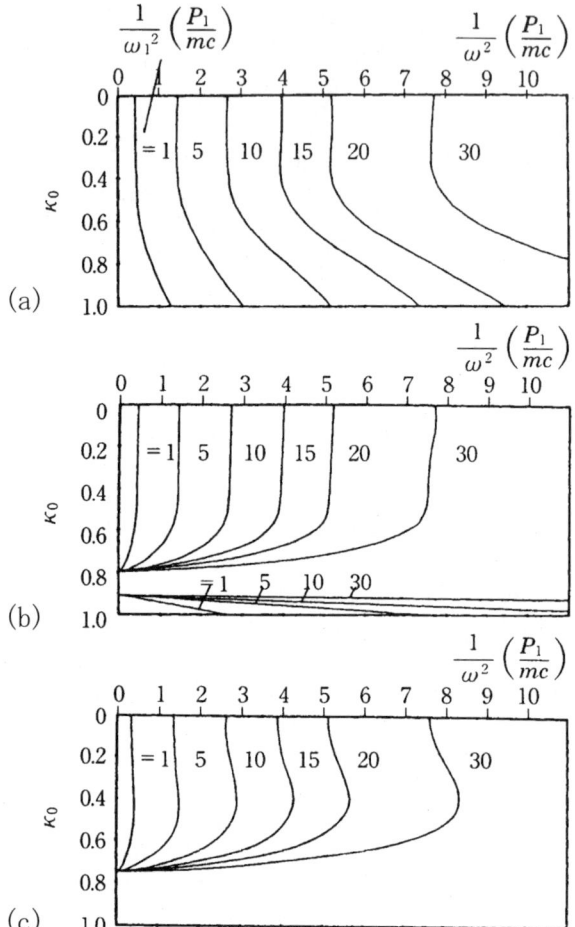

Fig. 5.38a-c. Stability charts for a shaft in porous bearings [36]. **a** $\Phi = 10^{-2}$, **b** $\Phi = 10^{-4}$, **c** $\Phi = 0$ (nonporous bush)

such as an earthquake. This was confirmed experimentally also by simulating an artificial miniearthquake with a hammer by hitting the pedestal which carries the rotating shaft (Adams [63]).

Further, if an imbalance and the bearing load of a rotor are changed variously while it is running in the hysteresis domain, it has been shown that all kinds of rotor responses can occur, e.g., periodic, quasiperiodic, and **chaotic** [63]. These examples are shown in Fig. 5.39. Which one of these three actually occurs and when the transition from one state to another occurs is very sensitively related to the operating conditions of the rotor. Thus, it has been reported that chaotic phenomena are useful

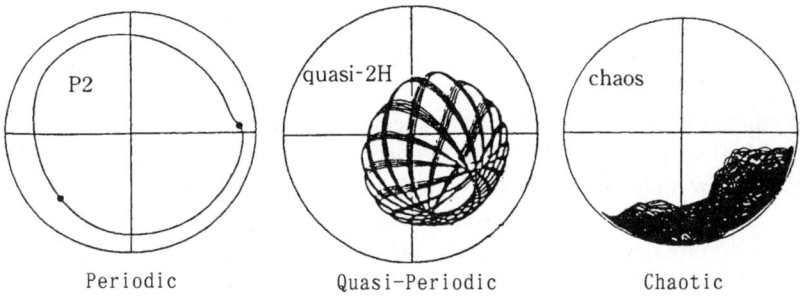

Fig. 5.39. Chaos in rotor-bearing system [63]

as safe diagnostic tools in assessing risks associated with the stable limit cycle within the hysteresis loop. [63].

5.10 Prevention of Oil Whip

It is very important to prevent oil whip in rotating machines. Some common methods are listed below:

1. Raise the critical speed of the shaft. The stable region in the stability chart is thereby expanded. Further, even if the shaft becomes unstable, if the rotating speed is lower than twice the critical speed, the shaft will not start a violent oil whip. The stable region can also be expanded by lowering the (length/diameter) ratio of the bearing.
2. Increase the eccentricity ratio of the journal (to reduce the floating height of the journal). The operation point of the shaft moves thereby into the stable region of the stability chart. The shaft is stable whenever the eccentricity ratio is larger than 0.8. To increase the eccentricity ratio, lowering the coefficient of viscosity of the oil, reducing the bearing length, increasing the bearing pressure, and increasing the bearing clearance are useful.

    While the above-mentioned methods are effective in circular bearings, the following methods using special bearings are also possible.
3. Use a noncircular bearing such as a two circular arc bearing or a three circular arc bearing. In these bearings, the radius of curvature of the metal surface is larger than that of a circular bearing, and hence the effective eccentricity ratio of the journal is large and so stability is high.
4. Use floating bush bearings. In this case, the bearing has two oil films, one inside and one outside the floating bush. Stability is generally improved by using a floating bush bearing, but the stability chart is complicated and sometimes stability can decrease. It is recommended that stability be examined for each case.
5. Use tilting pad bearings. In this case, the pad can tilt freely, and hence the coupling terms of the oil film coefficients are zero. Therefore, stability is essentially

high. The structure of the bearing is complicated, however, and hence the structural strength of the bearing can be low. It is suitable for low bearing pressures and high shaft speeds.

In summary, a circular bearing is adequate when bearing the pressure is high and the shaft speed is low. A tilting pad bearing is recommended when the bearing pressure is low and the shaft speed is high. A noncircular bearing or a floating bush bearing is used in intermediate cases. Many papers have been published on the stability of two circular arc and three circular arc bearings [40] [44], and also on the stability of tilting pad bearings [21] [27] [35] [46].

References

1. A. L. Kimball, "Internal Friction Theory of Shaft Whirling", *General Electric Review*, April, 1924, pp. 244 - 251.
2. B. L. Newkirk and H. D. Taylor, "Shaft Whipping Due to Oil Action in Journal Bearings", *General Electric Review*, August, 1925, pp. 559 - 568.
3. D. Robertson, "Whirling of a Journal in a Sleeve Bearing", *Philosophical Magazine*, Series 7, Vol. 15, January 1933, pp. 113 - 130.
4. H. Poritsky, "A Contribution to the Theory of Oil Whip", *Trans. ASME*, Vol. 75, 1953, pp. 1153 - 1161.
5. T. Okazaki, *Personal Communication*, 1954.
6. T. Okazaki and Y. Hori, "The Theory of Oil-Whip in Jouranl Bearings" (in Japanese), *Trans. JSME*, Vol. 21, No. 102, February 1955, pp. 125 - 130. Vol. 22, No. 114, February 1956, p. 90.
7. Y. Hori, "A Theory of Oil-Whip in Journal Bearings," *Proc. of the 5th Japan National Congress for Applied Mechanics*, September 7 - 9, Tokyo, 1955, pp. 395 - 398.
8. M. Hamanaka, *Personal Communication*, 1956.
9. B. L. Newkirk and J. F. Lewis, "Oil Film Whirl - An Investigation of Disturbances Due to Oil Films in Journal Bearings", *Trans. ASME*, Vol. 78, 1956, pp. 21 - 27.
10. B. L. Newkirk, "Varieties of Shaft Disturbances Due to Fluid Films in Journal Bearings", *Trans. ASME*, Vol. 78, 1956, pp. 985 - 988.
11. O. Pinkus, "Experimental Investigation of Resonant Whip", *Trans. ASME*, Vol. 78, 1956, pp. 975 - 983.
12. V.T. Morgan and A. Cameron, "Mechanism of Lubrication in Porous Bearings", *Proceedings of Conference on Lubrication and Wear*, IMechE, London, 1957, pp. 151 - 157.
13. T. Yamamoto, "On Critical Speeds of a Shaft Supported by a Ball Bearing", *Trans. ASME, Series E, J. of Applied Mechanics*, Vol. 26, No. 2, June 1959, pp. 199 - 204.
14. Y. Hori, "Oil Whip" (in Japanese), *Journal of JSME*, Vol. 61, No. 478, November 1958, pp. 1348 - 1356.
15. Y. Hori, "A Theory of Oil Whip", *Trans. ASME, Series E, J. of Applied Mechanics*, Vol. 26, No. 2, June 1959, pp. 189 - 198.
16. Y. Hori, "Study on Oil Whip" (in Japanese), Dissertation, University of Tokyo, December 1959.
17. T. Someya, "Stabilität einer in zylindrischen Gleitlagern laufenden, unwuchtfreien Welle", *Ingenieur-Archiv*, 33. Band, 2. Heft, 1963, pp. 85 - 108.

18. H. Gotoda, "Research on the Vibration of a Rotating Shaft Supported by Journal Bearings (in Japanese), Report of the Research Subcommittee on Radial Gasturbines, No. 4, JSME, July 15, 1963, pp. 114 - 206, Appendix pp. 207 - 244.
19. T. Someya, "Stability of a Balanced Shaft Running in Cylindrical Journal Bearings", *Proc. Lubrication and Wear Second Convention*, Eastbourne, Sponsored by IMechE, London, May 28-30, 1964, Paper 21, pp. 3 - 21.
20. H. Gotoda, "On the Vibration of a Rotating Shaft Supported by Journal Bearings (First Report, Unbalance Vibration)" (in Japanese), *Trans. JSME*, Vol. 30, No. 215, July 1964, pp. 887 - 900.
21. J.W. Lund, "Spring and Damping Coefficients for the Tilting Pad Journal Bearing", *ASLE Trans.*, Vol. 7, 1964, pp. 342 - 352.
22. M. Funakawa and A. Tatara, "Stability Criterion of an Elastic Rotor in Journal Bearings" (in Japanese), *Trans. JSME*, Vol. 30, No. 218, October 1964, pp. 1238 - 1244.
23. T. Someya, "Schwingungs- und Stabilitätsverhalten einer in zylindrischen Gleitlagern laufenden Welle mit Unwucht", *VDI - Forschungsheft 510*, 1965. pp. 1 - 36.
24. E. Nakagawa and H. Aoki, "Approximate Solution for Elastic and Damping Properties of Oil Film in Journal Bearings" (in Japanese), *Trans. JSME*, Vol. 31, No. 229, September 1965, pp. 1398 - 1408.
25. A. Cameron, "Principles of Lubrication", *Longman*, London, 1966
26. F.K. Orcutt and E.B. Arwas, "The Steady State and Dynamic Characteristics of a Full Circular Bearing and a Partial Arc Bearing in the Laminar and Turbulent Flow Regimes", *Trans. ASME, J. Lub. Tech.*, Vol. 89, July 1967, pp. 143 - 152.
27. F.K. Orcutt, "The Steady-State and Dynamic Characteristics of the Tilting-Pad Journal Bearing in Laminar and Turbulent Flow Regimes", *Trans. ASME, J. Lub. Tech.*, Vol. 89, July 1967, pp. 392 - 404.
28. K. Ono and A. Tamura, "On the Vibrations of a Horizontal Shaft Supported on Oil-Lubricated Journal Bearings" (in Japanese), *Trans. JSME*, Vol. 34, No. 258, February 1968, pp. 285 - 297.
29. T. Someya, "Dynamic Problems of Journal Bearing - Case Where It is Lubricated with Non-Compressive Liquid -" (in Japanesse), *Journal of JSME*, Vol 72, No. 610, Novwember 1969, pp. 1513 - 1523.
30. A. Tatara, "An Experimental Study on the Stabilizing Effect of Floating Bush Journal Bearings" (in Japanese), *Journal of JSME*, Vol. 72, No. 610, Novwember 1969, pp. 1564 - 1569.
31. M. Harada and H. Aoki, "The Dynamic Characteristics of Fully Circular Journal Bearings in the Turbulent Region" (in Japanese), *Journal of Japan Society of Lubrication Engineers*, Vol. 16, No. 6, June 1971, pp. 429 - 436.
32. K. Shiraki, "Troubleshooting of Vibration Problems in the Field" (in Japanese), *Journal of JSME*, Vol. 75, No. 639, April 1972, pp. 507 - 524.
33. M. Tanaka and Y. Hori, "Stability Characteristics of Floating Bush Bearings", *Trans. ASME*, Series F, Vol. 94, No. 3, July 1972, pp. 248 - 259.
34. G.L. Falkenhagen, E.J. Gunter and F.T. Schuller, "Stability abd Transient Motion of a Vertical Three-Lobe Bearing System", *J. Engineering for Industry, Trans. ASME*, Vol. 94, 1972, pp. 665 - 667.
35. S. Iida, "A Study on the Vibration Characteristics of Tilting Pad Journal Bearings" (in Japanese), *Trans. JSME*, Vol. 40, No. 331, March 1974, pp. 875 - 884.
36. Y. Hori and K. Okoshi, "Stability of a Rotating Shaft Supported by Porous Bearings", *Proc. the JSLE-ASLE International Lubrication Conference, Tokyo*, June 9 - 11, 1975, pp. 333 - 340.

37. A.G. Holmes, C.M.McC. Ettles and I.W. Mayes, "The Dynamics of Multi-Rotor Systems Supported on Oil Film Bearings", *Trans. ASME*, Series L, Vol. 100, No. 1, 1978, pp. 156 - 164.
38. K. Kikuchi, M. Takagi and S. Kobayashi, "Effect of Alignment on Vibration of Multi-Bearing Rotor System (First Report, Vibration Characteristics of Three-Bearing, One-Rotor System) (in Japanese), *Trans. JSME*, Vol. 45, No. 400, December 1979, pp. 1349 - 1356.
39. R. Uematsu and Y. Hori, "Influence of Misalignment of Support Journal Bearings on Stability of A Multi-Rotor System", *Tribology International*, Vol. 13, No. 5, Oct. 1980, pp. 249 - 252.
40. P.E. Allaire and R.D. Flack, "Journal Bearing Design for High Speed Turbomachinery", in *Bearing Design - Historical Aspects, Present Technology and Future Problems, ASME Publication*, Edited by W.J. Anderson, 1980, New York.
41. D.F. Li, K.C. Choy and P.E. Allaire, "Stability and Transient Characteristics of Four Multilobe Journal Bearing Configurations", *Journal of Luburication Technology, Trans. ASME*, Vol. 102, 1980, pp. 291 - 299.
42. T. Nasuda and Y. Hori, "Influence of Misalignment of Support Journal Bearings on Stability of Multi-Rotor Systems", *Proc. IFToMM Int. Conf. on Rotordynamic Problem in Power Plants, Rome*, September 28 - October 1, 1982, pp. 389 - 395.
43. M. Akkok and C.M.M. Ettles, "Journal Bearing Response to Excitation and Behaviour in the Unstable Region", *ASLE Trans.*, Vol. 27, 1984, pp. 341 - 351.
44. H. Hashimoto, S. Wada and H. Tsunoda, "Performance Characteristics of Elliptical Journal Bearings in Turbulent Flow Regime", *Bulletin, Japan Society of Mechanical Engineers*, 27 - 232, 1984, pp. 2265 - 2271.
45. M. Tanaka, K. Fukuda and Y. Hori, "Friction of Porous Metal Bearing" (in Japanese), *Journal of the Faculty of Engineering, University of Tokyo*, A, Vol. 23, 1985, pp. 16 - 17.
46. H. Hashimoto, S. Wada and T. Marukawa, "Performance Characteristics of Large Scale Tilting-Pad Journal Bearings", *Bulletin, Japan Society of Mechanical Engineers*, 28 - 242, 1985, pp. 1761 - 1768.
47. M. Malik and Y. Hori, "An Approximate Nonlinear Transient Analysis of Journal Bearing Response in Unstable Region of Linearized System", *Proc. IFToMM-JSME International Conference on Rotordynamics, Tokyo*, September 14 - 17, 1986, pp. 217 - 220.
48. Y. Hori, T. Kato, R. Wu and M. Tanaka, "Dynamic Characteristics of a Rotor of an Electric Motor for a Downhole Pump: A Vertical Flexible Rotor in a Three Lobe Stator", *Proc. International Conference on Mechanical Dynamics, Shengyang, China*, August 3 - 6, 1987, pp. 516 - 519.
49. T. Someya, Editor, "Journal Bearing Databook", *Springer Verlag*, Berlin Heidelberg, 1988
50. J. Mitsui, "Method of Calculation for Bearing Characteristics", in "Journal Bearing Databook", T. Someya, Editor, *Springer Verlag*, Berlin Heidelberg, 1988, pp. 231 - 240.
51. Y. Hori, "Anti-Earthquake Considerations in Rotordynamics", *Proc. 4th IMechE International Conference on Vibrations in Rotating Machinery*, Edinburgh, September 1988, pp. 1 - 8.
52. T. Kato and Y. Hori, "Application of the Matrix Form of the Reynolds Equation to Dynamic Journal Bearings" (in Japanese), *Trans. JSME*, C, Vol. 54, No. 500, April 1988, pp. 935 - 942.
53. T. Kato and Y. Hori, "Fast Method for Calculating Dynamic Coefficients of Finite Width Journal Bearings with Quasi-Reynolds Boundary Condition", *Trans. ASME. J. of Tribology*, Vol. 110, No. 3, July 1988, pp. 387 - 393.

54. T. Kato and Y. Hori, "On the Cross Terms of the Damping Coefficients of Finite Width Journal Bearings" (in Japanese), *Trans. JSME*, C, Vol. 54, No. 505, September 1988, pp. 2214-2217.
55. Y. Hori and T. Kato, "Seismic Effect on the Stability of a Rotor Supported by Oil Film Bearings" (in Japanese), *Trans. JSME*, C, Vol. 55, No. 511, March 1989, pp. 614-617.
56. T. Kato and Y. Hori, "Theoretical Condition for the $C_{xy} = C_{yx}$ Relation in Fluid Film Journal Bearings", *Trans. ASME, J. Tribology*, Vol. 111, No. 3, July 1989, pp. 426-429.
57. Y. Hori and T. Kato, "Earthquake-Induced Instability of a Rotor Supported by Oil Film Bearings", *Trans. ASME, J. Vibration and Acoustics*, Vol. 112, April 1990, pp. 160-165.
58. T. Kato, K. Koguchi and Y. Hori, "Seismic Response of a Multirotor System Supported by Oil Film Bearings" (in Japanese), *Trans. JSME*, C, Vol. 57, No. 544, December 1991, pp. 3761-3768.
59. T. Kato, H. Matsuoka and Y. Hori, "Seismic Response of a Multirotor System Supported by Oil Film Bearing (Effect of Misalignment)" (in Japanese), *Trans. JSME*, C, Vol. 58, No. 549, May 1992, pp. 1572-1578.
60. T. Kato, K. Koguchi and Y. Hori, "Seismic Response of a Linearly Stable Multirotor System", *Proc. IMechE International Conference on Vibrations in Rotating Machinery*, Bath, June. 1992, pp. 7-11.
61. T. Kato, H. Matsuoka and Y. Hori, "Seismic Response of a Linearly Stable, Misaligned Multirotor System", *Tribology Transaction, S.T.L.E.*, Vol. 36, 1993-2, pp. 311-315.
62. C.E. Koeneke, M. Tanaka and H. Motoi, "Axial Oil Film Rupture in High Speed Bearings Due to the Effect of the Centrifugal Force", *Trans. ASME, J. Tribology*, Vol. 117, No. 3, July 1995, pp. 394-398.
63. Maurice L. Adams, Michael L. Adams and J-S Guo, "Simulations and experiments of the non-linear hysteresis loop for rotor-bearing instability", *Proc. IMechE International Conference on Vibrations in Rotating Machinery*, September 9-12, Oxford, 1996, pp. 309-319.
64. K. Hatakenaka, M. Tanaka and K. Suzuki, "A Modified Reynolds Equation with Centrifugal Force being Considered and Its Application to Floating Bush Bearing" (in Japanese), *Trans. JSME*, C, Vol. 65, No. 636, August 1999, pp. 3395-3400.
65. K. Hatakenaka, M. Tanaka and K. Suzuki, "A Study on the Dynamic Characteristics of High Speed Journal Bearing with Axial Oil Film Rupture and Its Application o Stability Analysis of Floating Bush Bearing" (in Japanese), *Trans. JSME*, C, Vol. 65, No. 640, December 1999, pp. 4840-4845.
66. Y. Hori, Compilation of Manufacturers' Data from Various Sorces (2003).

6

Foil Bearings

A bearing surface is usually made so rigid that it will not deform under journal load or fluid film pressure. In some bearings, however, the bearing surface is made of foil or tape (metal foil or high polymer film, for example) that is sufficiently flexible. This kind of bearing is called a **foil bearing**. Figure 6.1a shows its fundamental form, where the angle β is called the **wrap angle**. Figure 6.1b, a combination of three basic units, is an example of a practical form of the foil bearing. Foil bearings were first studied by H. Blok and J. J. Van Rossum [1].

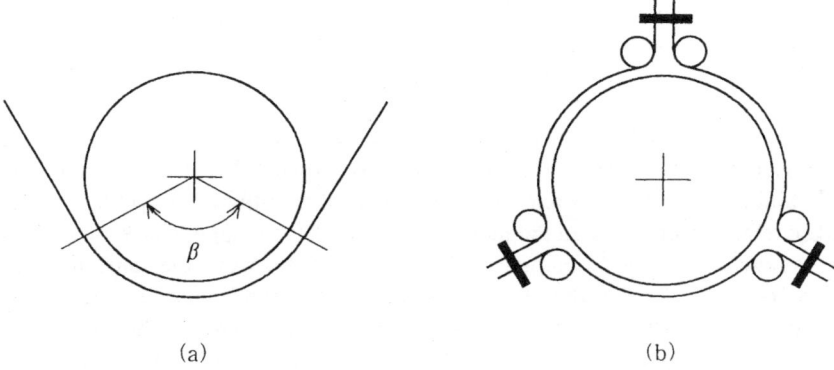

Fig. 6.1a,b. Foil bearings. **a** simple form, **b** combined form

In the case of a foil bearing, the foil deforms due to the pressure in the fluid film. If the foil deforms, the bearing clearance will naturally change and, in response to it, the pressure generated will also change. Thus, the foil shape (clearance shape) and the fluid film pressure interact closely with each other.

Figure 6.2 shows the distribution of pressure and clearance for a foil bearing Fig. 6.1a developed on a straight line. In a foil bearing, particularly when the wrap angle is large, it is known that the pressure and the clearance are almost constant over a quite wide range of the lubricating domain. Constancy of the pressure over a wide range of a foil bearing means that there is no particular force causing the shaft to whirl, and hence a shaft in a foil bearing has excellent stability [7]. Since the foil is not very strong mechanically, it can be said that a foil bearing is suitable to support a shaft with a low bearing load and low stability. A rotating shaft in an instrument used under zero-gravity conditions is a good example.

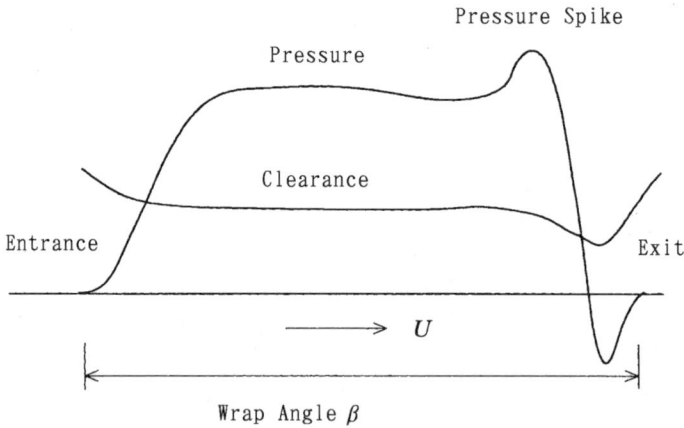

Fig. 6.2. Pressure and clearance of a foil bearing, showing a maximum in the pressure and a minimum in the clearance

Further, it is known that near the exit of the lubricating domain of a foil bearing, the pressure and the bearing clearance change as shown in Fig. 6.2, with a maximum and a minimum. A sharp increase in pressure is called a **pressure spike**.

The relationship between the magnetic head and the magnetic tape of a magnetic tape storage device for a computer is similar to that between the shaft and the foil of a foil bearing. The fact that the film thickness has a minimum near the exit of the lubricating domain is particularly important in this case. The smaller the clearance between the magnetic tape and the read/write element is, the higher the recording density can be. Therefore, a read/write element is installed at the minimum clearance position in magnetic tape storage devices. In this case, the surrounding air is automatically drawn into the space between the tape and the magnetic head and forms a fluid film.

In connection with magnetic tape storage, much research has been carried out into foil bearings [3]-[12]. In this section, a finite element method for a fluid film lubrication problem [13] is applied to a foil bearing, and the theoretical results are compared with experiments (Hori et al. [14] [15]). The profile of a lubricating surface of a magnetic head is often complicated, and the finite element method is suitable to the solution of such a problem.

6.1 Basic Equations

For a foil bearing, since the clearance distribution (foil shape) and the pressure distribution are mutually related, its analysis is mathematically a solution of the simultaneous equations of fluid film pressure and foil deformation. For simplicity, the following assumptions will be made:

1. Reynolds' equation is applicable to the fluid film.
2. Compressibility of the fluid can be disregarded.
3. The foil deforms easily.
4. Tension in the foil is constant regardless of time and location.
5. Flow and pressure in the fluid are uniform in the width direction of the foil tape.

Some notes should be added to these assumptions. In item (1), the extent of the domain in which Reynolds' equation can be applied is not clearly defined because the air in a very large space enters a gradually decreasing space and finally into a very thin clearance and then flows out again to the surroundings. Item (2) is valid when the pressure is low, but in some cases compressibility cannot be ignored. Item (3) does not hold in some cases where rigidity of the foil cannot be disregarded. Item (4) means that the viscosity of the fluid and the mass of the foil are very small. Item (5) means that the dimension of the lubricating domain in the flow direction is sufficiently small compared with that in the width direction of the tape.

On the above assumptions, the system will be described by the following simultaneous equations:

$$\frac{d}{dx}\left(\frac{h^3}{6\mu}\frac{dp}{dx}\right) = U\frac{dh}{dx} \tag{6.1}$$

$$p = T\left(\frac{1}{R} - \frac{d^2h}{dx^2}\right) \tag{6.2}$$

Equation 6.1 is Reynolds' equation for the fluid film where p is the fluid film pressure, h is the film thickness, μ is viscosity of the fluid, and x is the coordinate in the direction of foil movement. Equation 6.2, which is basically an equation of the balance of the fluid film pressure p and the foil tension T, shows the relation between pressure distribution p and film thickness distribution h in the fluid film on the assumption that the foil tension T is constant. R is the radius of the shaft.

The following boundary conditions are assumed (see Fig. 6.3):

1. At a point x_1 (the entrance of the lubricating domain), which is located upstream far enough but not too far from the entrance z_1 of the contact domain of the circular shaft and the foil at rest, it is assumed that pressure p is equal to the ambient pressure (i.e., zero) and that the clearance h is equal to h_1 when the foil is not moving.
2. At a point x_2 (the exit of the lubricating domain), which is located downstream far enough but not too far from the exit z_2 of the contact domain, as for the case of the entrance, it is assumed that pressure p is equal to the ambient pressure and that the clearance h is equal to h_2 when the foil is not moving.

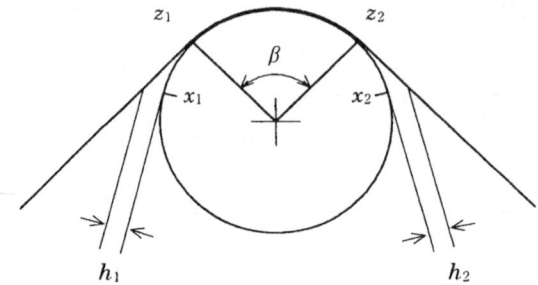

Fig. 6.3. Boundary conditions [15]

The locations of x_1, and x_2 are such that the film pressure generated can still be disregarded and the clearance is not so large that Reynolds' equation can still be used. It is difficult to determine the positions of x_1 and x_2 exactly, but it is expected that a little deviation from the exact position does not greatly affect the calculated results of pressure and clearance. Thus, the boundary conditions are as follows:

$$\begin{cases} p = 0 \text{ and } h = h_1 & \text{at } x = x_1 \\ p = 0 \text{ and } h = h_2 & \text{at } x = x_2 \end{cases} \quad (6.3)$$

6.2 Finite Element Solution of the Basic Equations

In solving the basic equations, Eqs. 6.1 and 6.2 under the boundary conditions Eq. 6.3, a **finite element method** is used [13]. It can be conveniently applied to a lubricating surface of complicated shape.

6.2.1 Reynolds' Equation

First, consider the following integral concerning the pressure distribution $p(x)$ over the interval (x_1, x_2):

$$J\{p\} = \int_{x_1}^{x_2} \left\{ \frac{h^3}{12\mu} \left(\frac{dp}{dx} \right)^2 - hU \frac{dp}{dx} \right\} dx \quad (6.4)$$

$J\{p\}$ is a function of the function $p(x)$ and is generally called a functional. When an arbitrary small change $\delta p(x)$ is given to the function $p(x)$, the first variation $\delta J\{p\}$ of $J\{p\}$ will be as follows:

$$\delta J\{p\} = -\int_{x_1}^{x_2} \left\{ \frac{d}{dx} \left(\frac{h^3}{6\mu} \frac{dp}{dx} \right) - U \frac{dh}{dx} \right\} \delta p \, dx \quad (6.5)$$

Equating this to zero yields the following stationary condition of the functional $J\{p\}$:

6.2 Finite Element Solution of the Basic Equations

$$\delta J\{p\} = -\int_{x_1}^{x_2} \left\{ \frac{d}{dx}\left(\frac{h^3}{6\mu}\frac{dp}{dx}\right) - U\frac{dh}{dx} \right\} \delta p \, dx = 0 \tag{6.6}$$

Since $\delta p(x)$ is an arbitrary function here, the stationary condition Eq. 6.6 is equivalent to Reynolds' equation, Eq. 6.1.

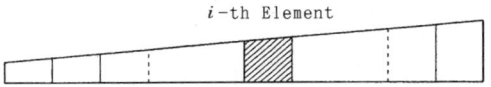

Fig. 6.4. Finite elements

To apply the finite element method to the problem, we divide the lubricating domain into N elements as shown in Fig. 6.4, and assume that the pressure in the ith element can be approximated by the following linear formula:

$$p_i(x) = \begin{bmatrix} \dfrac{x_{i+1} - x}{x_{i+1} - x_i}, & \dfrac{x - x_i}{x_{i+1} - x_i} \end{bmatrix} \begin{bmatrix} p_i \\ p_{i+1} \end{bmatrix} \tag{6.7}$$

where [ ] shows a matrix. This expression shows that the pressure at the ends (nodes) of the ith element are equal to p_i and p_{i+1}, respectively, and the pressure changes linearly between them. This can be written symbolically as follows:

$$p_i(x) = \overline{X_i} \cdot \overline{P_i} \tag{6.8}$$

With an augmented matrix X_i, the above expression can be written as:

$$p_i(x) = X_i \cdot P \tag{6.9}$$

where X_i and P are:

$$X_i = \begin{bmatrix} 0, 0, \cdots, \dfrac{x_{i+1} - x}{x_{i+1} - x_i}, \dfrac{x - x_i}{x_{i+1} - x_i}, \cdots, 0, 0 \end{bmatrix} \tag{6.10}$$

$$P = [p_1, p_2, \cdots\cdots, p_i, p_{i+1}, \cdots\cdots, p_{N-1}, p_N]^T \tag{6.11}$$

P is a column vector of all nodal pressures, and T in $[\ ]^T$ indicaters a transposed matrix.

Differentiating Eq. 6.9 with respect to x gives the following pressure gradient:

$$\frac{dp_i(x)}{dx} = R_i \cdot P \tag{6.12}$$

where R_i is:

$$R_i = \begin{bmatrix} 0, 0, \cdots, \dfrac{-1}{x_{i+1} - x_i}, \dfrac{1}{x_{i+1} - x_i}, \cdots, 0, 0 \end{bmatrix} \tag{6.13}$$

6 Foil Bearings

Then, the functional $J\{p\}$ of Eq. 6.4 can be reduced to:

$$J\{p\} = (P^T K_p - V) P \tag{6.14}$$

where K_p and V are:

$$K_p = \sum_{i=1}^{N} K_{pi} = \sum_{i=1}^{N} \int_{x_i}^{x_{i+1}} \frac{h_i^3}{12\mu} R_i^T R_i \, dx \tag{6.15}$$

$$V = \sum_{i=1}^{N} V_i = \sum_{i=1}^{N} \int_{x_i}^{x_{i+1}} h_i U R_i \, dx \tag{6.16}$$

Next, we equate the first variation of the functional of Eq. 6.14 to zero:

$$\delta J\{p\} = \sum_{i=1}^{N} \frac{\partial J}{\partial p_i} \delta p_i = 0 \tag{6.17}$$

Since δp_i is an arbitrary variable and K_p is symmetrical, the following relation is obtained:

$$\left\{ \frac{\partial J}{\partial p_i} \right\} = 2 K_p P - V = 0 \tag{6.18}$$

or

$$K_p P = \frac{1}{2} V \tag{6.19}$$

This is a matrix representation of the simultaneous linear equations for the nodal pressures P.

Now, let us approximate the film thickness of the ith element by the following linear equation:

$$h_i(x) = \left[\frac{x_{i+1} - x}{x_{i+1} - x_i}, \frac{x - x_i}{x_{i+1} - x_i} \right] \begin{bmatrix} h_i \\ h_{i+1} \end{bmatrix} \tag{6.20}$$

Then, Eqs. 6.15 and 6.16 give the following equations for the ith element:

$$\overline{K}_{pi} = \frac{1}{12\mu} \frac{1}{(x_{i+1} - x_i)^5} \left\{ \frac{1}{4}(h_{i+1} - h_i)^3 (x_{i+1}^4 - x_i^4) \right.$$
$$+ (h_{i+1} - h_i)^2 (h_i x_{i+1} - h_{i+1} x_i)(x_{i+1}^3 - x_i^3)$$
$$+ \frac{3}{2}(h_{i+1} - h_i)(h_i x_{i+1} - h_{i+1} x_i)^2 (x_{i+1}^2 - x_i^2)$$
$$\left. + (h_i x_{i+1} - h_{i+1} x_i)^3 (x_{i+1} - x_i) \right\} \begin{bmatrix} 1 & -1 \\ -1 & 1 \end{bmatrix} \tag{6.21}$$

$$\overline{V}_i = \frac{U}{(x_{i+1} - x_i)^2} \left\{ \frac{1}{2}(h_{i+1} - h_i)(x_{i+1}^2 - x_i^2) \right.$$
$$\left. + (h_i x_{i+1} - h_{i+1} x_i)(x_{i+1} - x_i) \right\} \begin{bmatrix} -1 & 1 \end{bmatrix} \tag{6.22}$$

where the bars over K_{pi} and V_i indicate that zeros in the augmented matrices were omitted. By using these matrices, it is possible to write down the matrix equation (Eq. 6.19) over all elements.

Therefore, if the nodal film thicknesses

$$H = [h_1, h_2, \cdots\cdots\cdots\cdots, h_N] \tag{6.23}$$

are given, the nodal film pressures

$$P = [p_1, p_2, \cdots\cdots\cdots\cdots, p_N] \tag{6.24}$$

can be determined by solving Eq. 6.19.

6.2.2 Equation of Balance for the Foil

Consider the following functional of the film thickness distribution $h(x)$ over the interval (x_1, x_2):

$$J\{h\} = \int_{x_1}^{x_2} \left\{ \frac{1}{2}\left(\frac{dh}{dx}\right)^2 - \left(\frac{p}{T} - \frac{1}{R}\right)h \right\} dx \tag{6.25}$$

Its first variation $\delta J\{h\}$ will be as follows (δh is an arbitrary small quantity):

$$\delta J\{h\} = -\int_{x_1}^{x_2} \left\{ \frac{d^2 h}{dx^2} + \left(\frac{p}{T} - \frac{1}{R}\right) \right\} \delta h\, dx \tag{6.26}$$

Equating this to zero gives the following stationary condition:

$$\delta J\{h\} = -\int_{x_1}^{x_2} \left\{ \frac{d^2 h}{dx^2} + \left(\frac{p}{T} - \frac{1}{R}\right) \right\} \delta h\, dx = 0 \tag{6.27}$$

which is equivalent to the equation of balance for the foil, Eq. 6.2, because δh is an arbitrary quantity. Therefore, the film thickness h is obtained by solving Eq. 6.27. The mathematical procedure hereafter is the same as that of the previous section.

If the film thickness in element $h_i(x)$ is approximated by Eq. 6.20, as before, the following equation will be obtained for the nodal film thicknesses H:

$$K_h H = W \tag{6.28}$$

where K_h and W are as follows:

$$\overline{K}_{hi} = \frac{1}{x_{i+1} - x_i} \begin{bmatrix} 1 & -1 \\ -1 & 1 \end{bmatrix} \tag{6.29}$$

$$\overline{W}_i = \frac{p_{i+1} - p_i}{T} \frac{1}{x_{i+1} - x_i} \left[\frac{1}{6} x_{i+1}^2 + \frac{1}{6} x_{i+1} x_i - \frac{1}{3} x_i^2, \right.$$

$$\left. \frac{1}{3} x_{i+1}^2 - \frac{1}{6} x_{i+1} x_i - \frac{1}{6} x_i^2 \right]$$

$$+ \left(\frac{p_i x_{i+1} - p_{i+1} x_i}{T} - \frac{x_{i+1} - x_i}{R} \right) \frac{1}{2} \begin{bmatrix} 1 & 1 \end{bmatrix} \tag{6.30}$$

where the bars over K_{hi} and W_i indicate that the zeros in the augmented matrix were omitted, as before.

6.2.3 Solution Procedure

Analysis of a foil bearing is thus a simultaneous problem composed of Eqs. 6.19 and 6.28. The two equations are given again here with new equation numbers:

$$K_p(H) P = \frac{1}{2} V(H) \tag{6.31}$$

$$K_h H = W(P) \tag{6.32}$$

The solution procedure is as follows. An appropriate film thickness distribution H_1 is first assumed, and the pressure distribution P_1 in that case is calculated by using Eq. 6.31. Then, the film thickness distribution H_2 for the pressure distribution P_1 is calculated by using Eq. 6.32. If:

$$\max \left| \frac{H_1 - H_2}{H_1} \right| < \epsilon \tag{6.33}$$

is satisfied for a sufficiently small quantity ϵ, H_1 will be the solution. If not, H_1 is modified in reference to H_2, and the same calculations are repeated until Eq. 6.33 is satisfied (an iterative method).

6.3 Characteristics of Foil Bearings

In this section, the pressure distribution and film thickness distribution for a single cylinder head and a double cylinder head, as shown in Fig. 6.5a,b, are calculated using the method of the previous section. Some of the calculated results are compared with experiments.

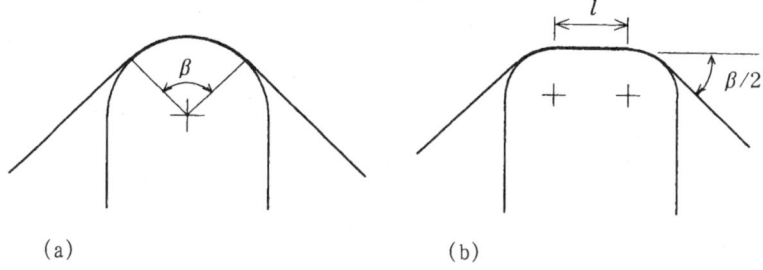

(a)   (b)

Fig. 6.5a,b. Single cylinder head (a) and double cylinder head (b)

It is assumed in the calculation that the entrance and the exit of the lubricating domain are located at a distance

$$x = R \left(\frac{6\mu U}{T} \right)^{(1/3)} \times (5.0 - 5.5) \tag{6.34}$$

from the entrance and the exit of the contact domain of the stationary foil in the upstream and the downstream directions, respectively. The reason for this is seen in Fig. 6.6.

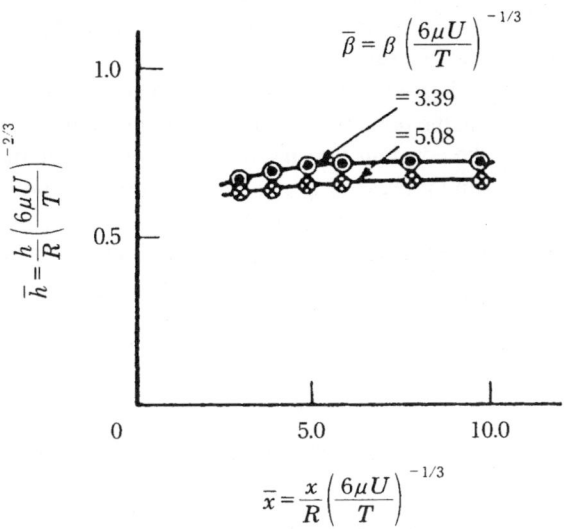

Fig. 6.6. Positions of the entrance and exit [15]

The figure shows the dependency of the calculated film thickness h at the center of the lubricating domain on the position x, where the boundary conditions are given, $\bar{h}$ and $\bar{x}$ being h and x nondimensionalized. Definitions of the nondimensional quantities are given in the figure. It is seen that $\bar{h}$ is nearly independent of $\bar{x}$ if $\bar{x} > 5$. Namely, the position $\bar{x} = 5$ is considered to be "far enough." And this is "not too far." Because calculaton shows that Reynolds' number at the position $\bar{x} = 5$ is less than 500, then Reynolds' equation, Eq. 6.1, can be used there. Therefore, $\bar{x} = 5$ is a suitable location for the boundary conditions. $\bar{\beta}$ in the figure is the nondimensional wrap angle.

6.3.1 Single Cylinder Heads

Figure 6.7 shows the nondimensional pressure distribution $\bar{p}$ and the nondimensional film thickness distribution $\bar{h}$ of a single cylinder head (cf. Fig. 6.5a,ba) for small, intermediate, and large wrap angles $\bar{\beta}$. Definitions of $\bar{p}$ and $\bar{h}$ are given in the figure together with that of $\bar{\beta}$. Toward the end of the transition from Fig. 6.7a to Fig. 6.7b, the region of constant film thickness and that of constant pressure begin to appear. In Fig. 6.7c, wide domains of constant film thickness and constant pressure are clearly seen; the minimum film thickness appears near the exit and, corresponding to this, the maximum pressure (pressure spike) and the minimum pressure (pressure valley) appear before and after the point of minimum film thickness. These phenomena near the exit are well known as the exit effects of a foil bearing. In the case of a magnetic tape memory storage device, the read/write element is installed near the point of minimum film thickness in Fig. 6.7c, as stated before. It is known that the recording density goes up in almost inverse proportion to the size of the clearance between the read/write element and the recording surface.

Figure 6.8 shows the dependence of the nondimensional constant film thickness $\overline{h}^*$ (the film thickness h^* at the point where the pressure gradient becomes zero for the first time is defined as the constant film thickness, cf. Fig. 6.7c), the minimum film thickness $\overline{h}_{min}$, the maximum pressure $\overline{p}_{max}$, and the minimum pressure $\overline{p}_{min}$ on the nondimensional wrap angle $\overline{\beta}$. As seen in the figure, these values are almost constant for $\overline{\beta} > 5$. The constant film thickness h^* and the minimum film thickness h_{min} are formulated as follows from the figure.

$$h^* \approx 0.64R\left(\frac{6\mu U}{T}\right)^{2/3} \tag{6.35}$$

$$h_{min} \approx 0.44R\left(\frac{6\mu U}{T}\right)^{2/3} \tag{6.36}$$

Figure 6.9 shows the dependencies of the positions $\overline{x}$ of $\overline{h}_{min}$, $\overline{p}_{min}$, and $\overline{p}_{max}$ on the nondimensional wrap angle $\overline{\beta}$. These positions are measured from the geometric point of contact z_2 (see Fig. 6.3). It is seen from the figure that these positions are nearly independent of $\overline{\beta}$ when $\overline{\beta} > 5$. This means that, if $\overline{\beta} > 5$, the foil shape in the exit domain hardly changes. This is true in the entrance domain also. Therefore, it is seen that, if $\overline{\beta} > 5$, even if $\overline{\beta}$ becomes large, the foil shape in the entrance and the exit domains does not change but the domain of constant film thickness is simply extended (cf. Fig. 6.6).

6.3.2 Double Cylinder Heads

Figure 6.10 shows the distributions of the fluid film pressure $\overline{p}$ and the film thickness $\overline{h}$ in the case of a double cylinder head of wrap angle $\overline{\beta} = 2.48$. The solid line in the figure corresponds to the case $\overline{l} = 1.93$, where $\overline{l}$ is the nondimensional length of the flat part connecting the two cylinders and the dashed line is the case $\overline{l} = 0$. The latter is equivalent to a single cylinder head. For $\overline{l} = 1.93$, two maxima of pressure and two minima of film thickness are seen, corresponding to the two cylinders. In the case of a magnetic tape memory storage device, a read/write element is installed at the position of the minimum film thickness. The exit effect, which is clear in the case of a single cylinder head, is not clearly seen except for the pressure valley just behind the exit.

In the flat part, the pressure is lower and the film thickness is larger than those in the cylindrical part.

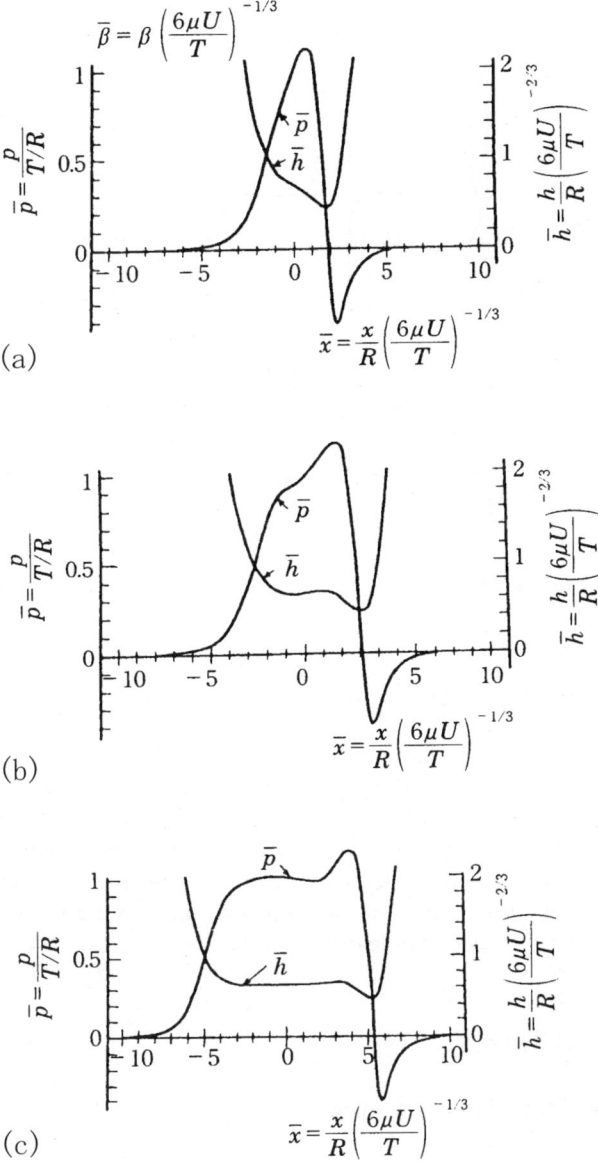

Fig. 6.7a–c. Pressure and clearance distribution in a single cylinder head [15]. **a** $\bar{\beta} = 2.48$, **b** $\bar{\beta} = 5.08$, **c** $\bar{\beta} = 9.53$

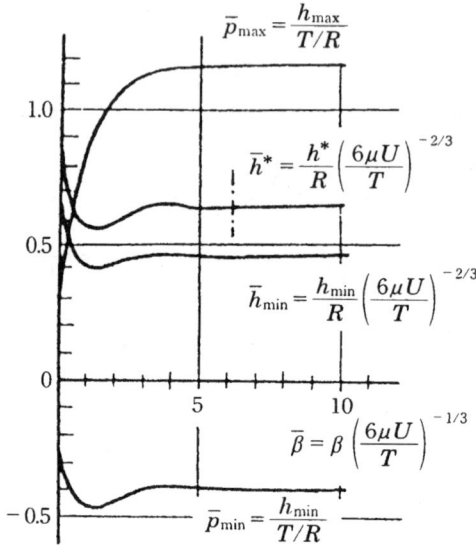

Fig. 6.8. Dependence of $\bar{h}^*$, $\bar{h}_{\min}$, $\bar{p}_{\max}$, and $\bar{p}_{\min}$ on $\bar{\beta}$ [15]

6.3.3 Comparison with Experiments

Figure 6.11a,b shows comparisons between experimental and the theoretical results of the pressure distribution $\bar{p}$ and the film thickness distribution $\bar{h}$ in single cylinder heads for two different wrap angles. In both cases, the experiments and the calculations are in fairly good agreement, and the constant film thickness domain and the pressure spike are actually observed. The air film thickness was measured using the electric capacity and the air film pressure was measured with a semiconductor pressure gauge through a small hole drilled in the electrode for the capacity.

6.4 Additional Topics

6.4.1 Magnetic Tape Memory Storage

In a magnetic tape memory storage device, magnetic tape runs over a magnetic head with a very thin air film between them, and a read/write element is installed in the magnetic head. In this case, it is necessary to decrease the clearance between the magnetic tape and the read/write element to obtain high density of recording, therefore much effort has been made to make the air film between magnetic tape and magnetic head as thin as possible. In recent years, an air film thickness of about $h = 100$ nm has been achieved, i.e., of the order of the mean free path of an air molecule $\lambda = 70$ nm. Therefore, the air cannot be regarded as a continuum but shows

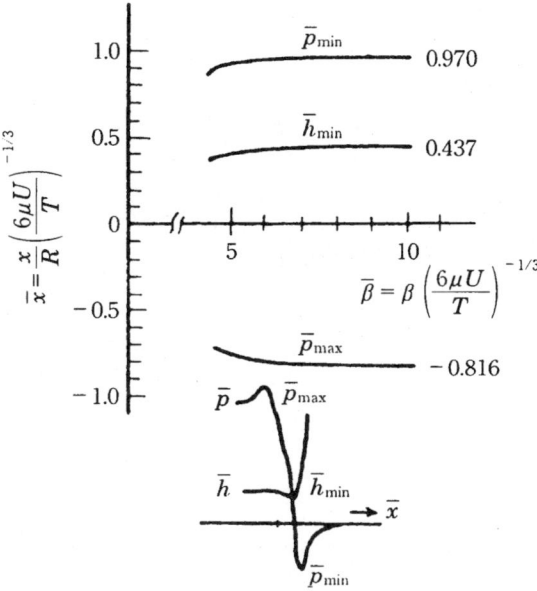

Fig. 6.9. Dependencies of the positions of $\bar{h}_{\min}$, $\bar{p}_{\min}$, and $\bar{p}_{\max}$ on $\bar{\beta}$ [15]

the particulate nature of the molecules. The effect of the particulate nature appears as slip between the solid surface and the air (slip flow) and equivalently as a decrease in viscosity.

The modified Reynolds' equation considering slip flow was derived by Burgdorfer [2] (cf. Section 4.4.2) as follows:

$$\frac{\partial}{\partial x}\left[ph^3\frac{\partial p}{\partial x}\left(1 + 6\frac{\lambda}{h}\right)\right] + \frac{\partial}{\partial y}\left[ph^3\frac{\partial p}{\partial y}\left(1 + 6\frac{\lambda}{h}\right)\right] = 6\mu U\frac{\partial(ph)}{\partial x} \qquad (6.37)$$

This equation was derived for an air journal bearing, and is considered to be applicable when $0 < \lambda/h \ll 1$, where λ/h is called the Knudsen number and is often written as M.

Although the condition of $\lambda/h \ll 1$ is not satisfied in recent magnetic tape memory storage devices, it is understood that, as previously stated (Section 4.4.2), the above equation can in fact be used for an air film thickness of $h = 100$ nm.

Throughout this chapter, foil bearings were analyzed with the usual Reynolds' equation. This may lead to an air film several percent thicker than the actual film for the very thin films in recent magnetic tape memory storage devices.

6.4.2 Foil Disk

A **foil disk** is a sheet of circular foil, a typical example of which is a floppy disk for a personal computer. If several foil disks, parallel to each other, are attached to a

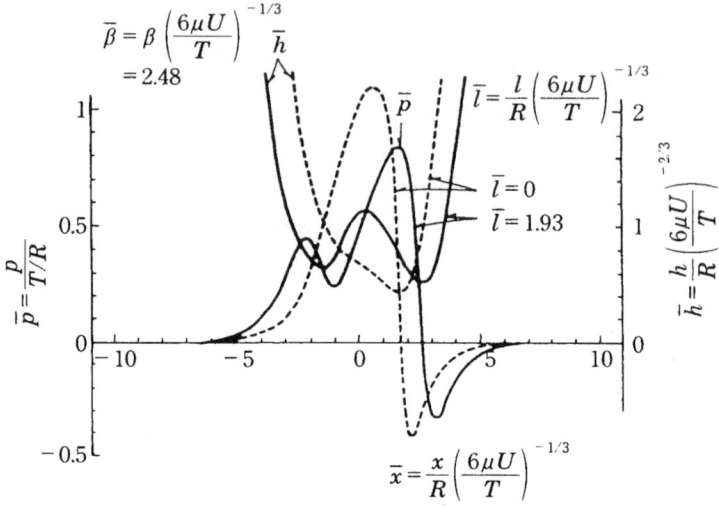

Fig. 6.10. Distributions of pressure and clearance in a double cylinder head [14]

rotating axle at right angles, a kind of secondary flow arises in the air layer between adjacent disks. If a foil disk attached to the end of a rotating axle rotates parallel to a rigid plane wall, another kind of secondary flow arises in the air layer between the foil disk and the rigid wall. In these cases, a complicated and interesting wave (out-of-plane deformation) appears in the foil as a result of the secondary air flow. An experimental study of the latter phenomenon is described here (Hasuike et al. [16]).

An experimental rig is outlined in Fig. 6.12. In order to observe the foil deformation by the moiré method (moiré is the wave pattern), a grating glass (a glass plate with many fine parallel grooves) is used as the rigid plane. The size of the grating glass is 200 mm × 200 mm × 10 mm, the line density of the grooves is 200/inch, and the width of a groove is equal to that of the transparent part between the grooves. The foil disk is placed face to face near the grated surface of the glass plate and is rotated by a variable speed motor. The foil disk is a circular sheet of polyester film 200 mm in diameter and 83 μm thick, and is attached to a rotating axle through a flange of diameter 54 mm. The rotating speed is variable in the range 300–3000 rpm. The rotating axle for the foil disk is hollow, and air can be blown out between the glass plate and the foil disk, if needed.

A beam of light is projected onto the grating glass with an incident angle of 45° to the glass surface and at 90° to the grating lines to form a moiré pattern by interference of the shadow of the grating on the foil and the grating itself. The moiré pattern in this case can be interpreted as contour lines of the foil warping, and the difference of height of the foil between adjacent fringes is equal to the pitch of the grating, or approximately 127 μm. A hot wire anemometer is placed near the periphery of the foil disk to observe the air flow out of the foil disk. A tungsten wire 5 μm in diameter

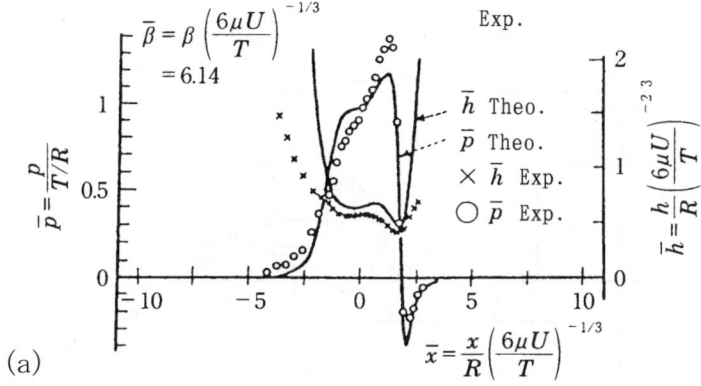

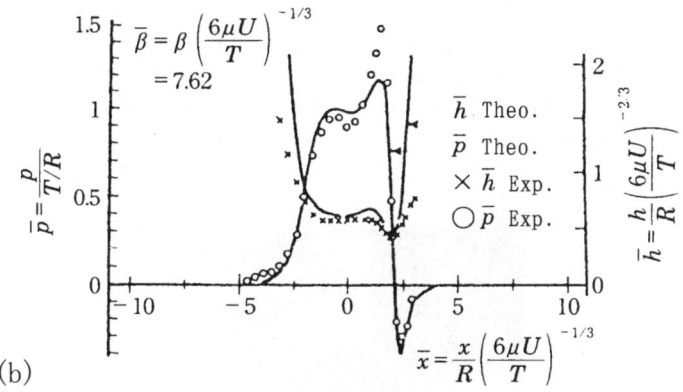

Fig. 6.11a,b. Comparisons of calculated results and experiments (single cylinder head) [15]. **a** foil width 2.54 cm, head radius 5.0 cm, wrap angle 8°, foil velocity 15.7 m/s, foil tension 0.167 kg/cm. **b** wrap angle 10°, the other parameters being the same as those in **a**.

and 1.2 mm in length is stretched in a pipe (copper) with an entrance diameter of 6 mm and a length of 35 mm.

Figure 6.13a shows an example of the moiré pattern thus obtained. The experimental conditions were: foil clearance $h = 2$ mm at the position of the axle, blow-out air pressure $p_{in} = 0$ Pa, and rotational speed $\Omega = 314$ rad/s. Since the moiré pattern is equivalent to contour lines, this figure shows that a domain where the foil is lifted from the glass extends along radii from the center of the disk to the upper left and to the lower right. The clearance between the foil and the glass plate is large at these

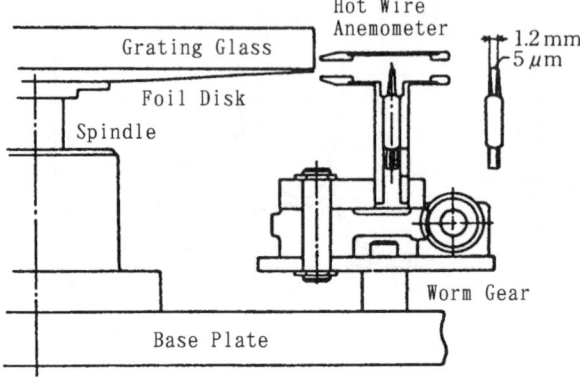

Fig. 6.12. Experimental apparatus for a foil disk [16]

locations, forming a tunnel-like space. In this domain, the air can flow easily and actually the air flows outward in the radial direction from the center of the disk as a result of centrifugal force. In the area between the tunnels where the clearance is small, the air flows inward slowly.

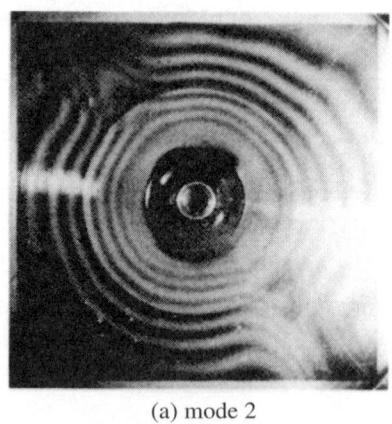

(a) mode 2

Fig. 6.13a. A moiré pattern on a foil disk (1) [16]

Since there are two tunnel-like areas in this case, it is designated as mode 2. Figure 6.13b-d shows mode 3, mode 4 and mode 5 configurations for the same values of $h = 2$ mm and $p_{in} = 0$ Pa as above and the value of $\Omega = 147$ rad/s. Here, it is interpreted from the moiré pattern that 3, 4, and 5 tunnels are formed. Since several kinds of moiré pattern are observed under the same experimental conditions

6.4 Additional Topics   135

(the same foil rotating speed, the same clearance, and so forth), this phenomenon is considered to be a kind of eigenvalue problem. Various modes can be obtained by giving small disturbances to foil disks by air jets, for example.

Figure 6.13e-g shows mode 6, 7, and 8 patterns found for only slightly different experimental conditions.

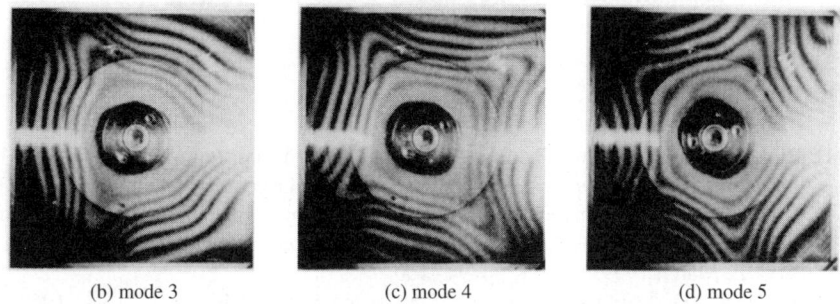

(b) mode 3   (c) mode 4   (d) mode 5

Fig. 6.13b-d. Moire patterns on a foil disk (2) [16]. **b** mode 3, **c** mode 4, **d** mode 5

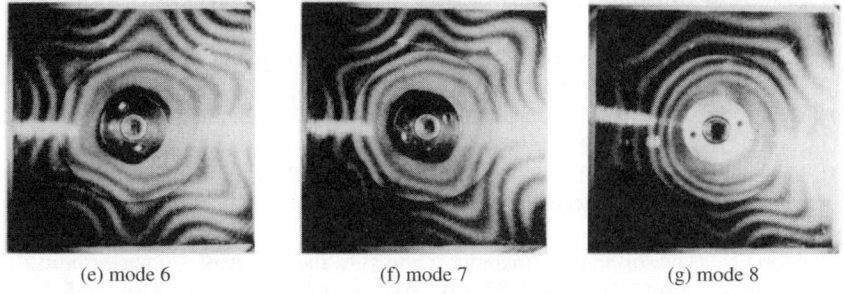

(e) mode 6   (f) mode 7   (g) mode 8

Fig. 6.13e-g. Moire patterns on a foil disk (3) [16]. **e** mode 6, **f** mode 7, **g** mode 8

These photographs were taken using a stroboscopic technique with synchronization to the rotation of the moiré pattern, and, from the flash speed of the stroboscope when the moiré came to a standstill, it was found that the moiré patterns was rotating a little slower than half the rotating speed of the foil disk. That is, the tunnel-like warping of the foil disk is rotating not at the speed of the foil disk but a significantly lower speed. The explanation for this is that the average flow velocities of the air between the disk and the glass plate in the circumferential direction is approximately one-half of that of the disk speed at the same point, and the disk warping is moving like a wave on the air flow over the disk surface. The hot wire anemometer showed that the air was blowing out in a jet from the tunnel portion of the disk foil and the frequency of the air jet also showed that the rotational speed of the tunnel was half that of the foil disk.

It is reported also in the case of foil bearings that a bump made by a disturbance in the foil runs at approximately half the running speed of the foil [6]. This can also be explained from the fact that the average flow velocity of the air between the head and the foil is approximately half the running speed of the foil.

References

1. H. Blok, J.J. Van Rossum, "The Foil Bearing - A New Departure in Hydrodynamic Lubrication", *Lubrication Engineering*, Vol. 9, No. 6, December, 1953, pp. 316-320.
2. A. Burgdorfer, "The Influence of the Molecular Mean Free Path on the Performance of Hydrodynamic Gas Lubricated Bearings", *Journal of Basic Engineering, Trans. ASME*, March 1959, Vol. 81, pp. 94-100.
3. W.A. Gross, "Gas Film Lubrication", *John Wiley & Sons, Inc., New York*, 1962, pp. 138-141.
4. E.J. Barlow, "Self-Acting Foil Bearings of Infinite Width", *Trans. ASME. F*, Vol. 89, No. 3, July 1967, pp. 341-345.
5. M. Wildman, "Foil Bearings", *Trans. ASME. F*, Vol. 91, No. 1, January 1969, pp. 37-44.
6. A. Eshel, "The Propagation of Disturbances in the Infinitely Wide Foil Bearing", *Trans. ASME. F*, Vol. 91, No. 1, January 1969, pp. 120-125.
7. L. Licht, "An Experimental Study of High Speed Rotors Supported by Air-Lubricated Foil Bearings, Part I & II", *Trans. ASME. F*, Vol. 91, No. 3, July 1969, pp. 477-505.
8. A. Eshel, "On Controlling the Film Thickness in Self-Acting Foil Bearings", *Trans. ASME. F*, Vol. 92, No. 2, April 1970, pp. 359-362.
9. A. Eshel, "On Fluid Inertia Effects in Infinitely Wide Foil Bearings", *Trans. ASME. F*, Vol. 92, No. 3, July 1970, pp. 490-494.
10. H. Mori, K. Hayashi and T. Yokomi, "A Study on Foil Bearing (An Experimental Investigation of Foil Displacement)" (in Japanese), *Trans. JSME*, Vol. 37, No. 295, March 1971, pp. 602-610.
11. H. Mori, K. Hayashi and T. Yokomi, "Ditto (On the Effect of Compressibility)" (in Japanese), *Trans. JSME*, Vol. 37, No. 303, November 1971, pp. 2229-2235.
12. T. Barnum, H.G. Elrod, Jr., "An Experimental Study of the Dynamic Behavior of Foil Bearings", *Trans. ASME. F*, Vol. 94, No. 1, January 1972, pp. 93-100.
13. M.M. Reddi, "Finite Element Solution of the Incompressible Lubrication Problem", *Trans. ASME. F*, Vol. 91, No. 3, July 1969, pp. 524-533.
14. Y. Hori, A. Hasuike, T. Higashi and Y. Nagase, "A Study on Foil Bearing" (in Japanese), Journal of Faculty of Engineering, University of Tokyo, A-12, 1974, pp. 16-17.
15. Y. Hori, A. Hasuike, T. Higashi, Y. Nagase, "A Study on Foil Bearings - An Application to Tape Memory Devices -", *Proc of 1975 Joint ASME-JSME Applied Mechanics Western Conference, Honolulu, Hawaii*, March 24-27, 1975, pp. 121-125 #D-5. *Bulletin of the JSME* 20-141 (1977-3) pp. 381-387.
16. A. Hasuike and Y. Hori, "A Study on Foil Disk" (in Japanese), Trans. JSME, C, Vol. 49, No. 440, April 1983, pp. 704-707.

7
Squeeze Film

Pressure arises in a fluid film between two mutually approaching surfaces. This is called the **squeeze effect** and the fluid film is called the **squeeze film**. O. Reynolds referred to the squeeze effect in his famous paper on lubrication (1886) and stated that it was an important mechanism, together with the wedge effect, for the generation of pressure in a lubricating film. Especially when a sufficiently large wedge effect is not expected, for example in the case of the small-end bearing of a crank for a reciprocating engine or in the case of an animal joint, he wrote that the squeeze film effect was the only mechanism for pressure generation. It is surprising that the lubrication mechanism of animal joints was discussed over 100 years ago. The fact that the rubber sole of a shoe or a rubber tire on a car is very slippery on a wet road surface can be understood as a similar phenomenon. In this case a thin water film hinders the contact of the rubber and the road surface.

In the above examples, two mutually approaching surfaces were considered, however, two mutually receding surfaces are also worth considering. In this case, since negative pressure arises in the fluid film, this phenomenon is called negative squeeze. The case of two approaching surfaces is called positive squeeze.

Further, it is also interesting to consider situations in which positive and negative squeeze occur alternately. In the small-end bearing of a crank and in an animal joint, a positive and a negative load acts by turns, and positive and negative squeeze occurs alternately. In this case, fluid is sucked into the gap between the two surfaces during the negative squeeze (negative pressure arises) and the fluid is squeezed out during the positive squeeze (positive pressure arises) and supports a load. It is interesting that, even when the positive and negative movement of the two surfaces is perfectly symmetrical, a positive load capability arises in many cases on balance through various mechanisms, as will be seen later. This phenomenon is a form of rectification.

A squeeze film is, unlike a wedge film, always in an unsteady state. Even when the added load is constant, a squeeze film becomes either thinner gradually with time or thicker, and is never in a stationary state except for the case of zero load. Therefore, a squeeze film cannot be maintained for a long time under a constant load, but is maintained for a long time only when positive and negative squeezes are repeated alternately.

7.1 Basic Equations

As preparation for dealing with a squeeze film between two disks, the basic equations of a squeeze film in cylindrical coordinates (r, θ, z) will be introduced (Kuroda et al. [7]).

a. Navier–Stokes Equation

When a phenomenon is axisymmetric and ρ and μ are constant, the Navier–Stokes equations in cylindrical coordinates (r, θ, z) are written as follows:

$$\rho\left(\frac{\partial v_r}{\partial t} + v_r\frac{\partial v_r}{\partial r} + v_z\frac{\partial v_r}{\partial z}\right) = -\frac{\partial p}{\partial r} + \mu\left(\frac{\partial^2 v_r}{\partial r^2} + \frac{1}{r}\frac{\partial v_r}{\partial r} + \frac{\partial^2 v_r}{\partial z^2} - \frac{v_r}{r^2}\right) \quad (7.1)$$

$$\rho\left(\frac{\partial v_z}{\partial t} + v_r\frac{\partial v_z}{\partial r} + v_z\frac{\partial v_z}{\partial z}\right) = -\frac{\partial p}{\partial z} + \mu\left(\frac{\partial^2 v_z}{\partial r^2} + \frac{1}{r}\frac{\partial v_z}{\partial r} + \frac{\partial^2 v_z}{\partial z^2}\right) \quad (7.2)$$

where v_r and v_z are the fluid velocity in the radial and the axial direction, respectively.

In Fig. 7.1, it is assumed that the film thickness h is sufficiently small compared with the radius of the squeeze surface r_a, i.e., $h \ll r_a$. In this case, a comparison of the order of magnitude of the above two equations gives $\frac{\partial p}{\partial r} \gg \frac{\partial p}{\partial z}$, therefore only Eq. 7.1 will be considered hereafter. If $h \ll r_a$, Eq. 7.1 will be as follows:

$$\rho\left(\frac{\partial v_r}{\partial t} + v_r\frac{\partial v_r}{\partial r} + v_z\frac{\partial v_r}{\partial z}\right) = -\frac{\partial p}{\partial r} + \mu\frac{\partial^2 v_r}{\partial z^2} \quad (7.3)$$

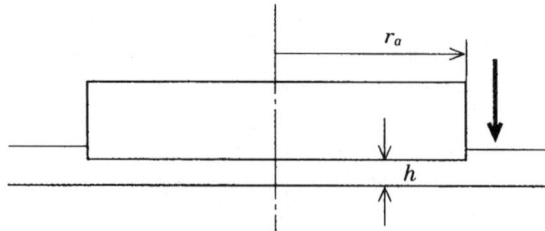

Fig. 7.1. Squeeze film

b. Continuity Equation

The continuity equation in cylindrical coordinates is:

$$\frac{1}{r}\frac{\partial}{\partial r}(rv_r) + \frac{\partial v_z}{\partial z} = 0 \tag{7.4}$$

The equation for a squeeze motion can be written as:

$$2\pi r \int_0^h v_r dz = -\pi r^2 \dot{h} \quad \text{(rigid surface)} \tag{7.5}$$

$$= -2\pi \int_0^r r\dot{h} dr \quad \text{(soft surface)} \tag{7.6}$$

where $\dot{h} = \partial h/\partial t$ is the relative velocity of the two surfaces (note that $\dot{h} < 0$ for a positive squeeze and $\dot{h} > 0$ for a negative squeeze).

An analysis of a squeeze film including inertia effects can be performed using three equations: Eqs. 7.3, 7.4, and 7.5 (or Eq. 7.6).

c. Reynolds' Equation

When inertia effects can be disregarded, Reynolds' equation can be derived. First, simplify the Navier–Stokes equation, Eq. 7.3, as follows:

$$\frac{\partial p}{\partial r} = \mu \frac{\partial^2 v_r}{\partial z^2} \tag{7.7}$$

Integration of the above equation twice with respect to z under the boundary condition $v_r = 0$ at $z = 0$ and $z = h$ gives the flow velocity v_r as follows:

$$v_r = \frac{1}{2\mu}\frac{\partial p}{\partial r}(z^2 - hz) \tag{7.8}$$

Substituting this into the continuity equation, Eq. 7.4, and integrating that with respect to z from 0 to h under the boundary condition $v_z = 0$ at $z = 0$, $v_z = \dot{h}$ at $z = h$ yields Reynolds' equation in cylindrical coordinates as follows:

$$\frac{\partial}{\partial r}\left(rh^3\frac{\partial p}{\partial r}\right) = 12\mu r\dot{h} \tag{7.9}$$

d. Boundary Conditions for Pressure

If the fluid inertia can be neglected, the pressure at the periphery of the squeeze film is equal to the ambient pressure (i.e., zero). Therefore, the boundary condition will be:

$$p = 0 \quad \text{at} \quad r = r_a \tag{7.10}$$

If the fluid inertia is taken into consideration, the boundary conditions for a positive squeeze and that for a negative squeeze are different, and are as follows, respectively:

If $\dot{h} < 0$, $p = 0$ at $r = r_a$ (7.11)

If $\dot{h} > 0$, $p = -\Delta p$ at $r = r_a$ (7.12)

Whereas for a positive squeeze (Eq. 7.11), the pressure at the periphery of the squeeze film is equal to the ambient pressure (i.e., zero), in the case of a negative squeeze (Eq. 7.12), a pressure drop $-\Delta p$ occurs when the fluid is sucked into the gap between disks, and the pressure at the periphery of the squeeze film becomes lower than the ambient pressure by the amount Δp.

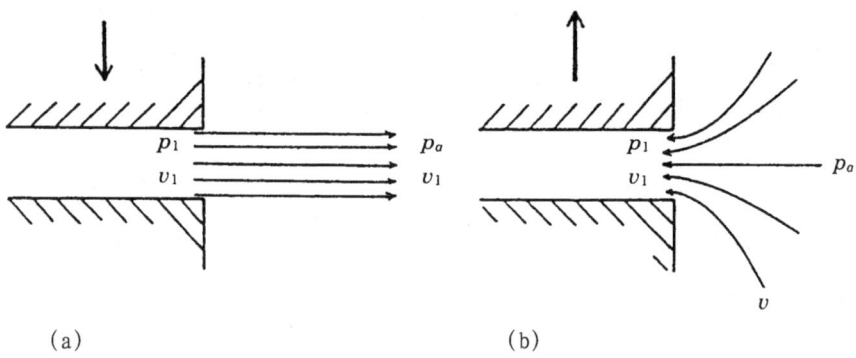

Fig. 7.2a,b. Boundary condition in a squeeze film [7]. **a** positive squeeze, **b** negative squeeze

This is clearly seen in Fig. 7.2a,b. For positive squeeze, the fluid is squeezed out as a jet as shown in Fig. 7.2a,ba, and there is no difference in the flow velocity inside and outside the edge of the disk ($r = r_a - 0$ and $r = r_a + 0$). Therefore, there is no difference in pressure either, from Bernoulli's equation. Therefore, Eq. 7.11 can be used as a boundary condition (pressure at $r = r_a - 0$). In constrast, for negative squeeze, the surrounding fluid is sucked into the gap between the disks along the streamlines shown Fig. 7.2a,bb, and the fluid is contracted rapidly when entering the gap between the disks. Therefore, the flow velocity increases rapidly and a pressure drop takes place. Now, consider an ideal fluid for simplicity, and let the pressure be zero and the flow velocity also be zero outside the disks, and let the pressure be p_1 and the flow velocity be v_1 just inside the gap between the disks, then Bernoulli's equation

$$p_1 + \frac{1}{2}\rho v_1^2 = 0 + 0 \qquad (7.13)$$

gives p_1 as follows:

$$p_1 = -\frac{1}{2}\rho v_1^2 = -\frac{1}{8}\rho r_a^2 \left(\frac{\dot{h}}{h}\right)^2 \qquad (7.14)$$

where $\dot{h}$ is the mutual receding velocity of the disks. The pressure drop Δp will be:

$$\Delta p = \frac{1}{8}\rho r_a^2 \left(\frac{\dot{h}}{h}\right)^2 \tag{7.15}$$

Actually, the flow pattern at the entrance to the gap between the disks is complicated, and the value of Δp will change with various factors, including the roundness of the edge of the disk. There is an empirical formula which gives a pressure drop of double the above-mentioned value in the case of a sharp edge, because the flow is contracted by fluid inertia. i.e.,

$$\Delta p = \frac{1}{4}\rho r_a^2 \left(\frac{\dot{h}}{h}\right)^2 \tag{7.16}$$

7.2 Squeeze Between Rigid Surfaces

The basic issues of a squeeze between rigid surfaces will be considered first (Kuroda et al. [7]).

7.2.1 Squeeze Without Fluid Inertia

Let the squeeze surfaces be rigid, the squeezing velocity be sufficiently small, and the fluid inertia be neglected. Let the radius of the disk be r_a, the gap between the disks be h, the fluid velocity in the radial direction be $v_r(z, r, t)$, that in the film thickness direction be $v_z(z, t)$, and the fluid pressure be $p(r, t)$.

As a basic equation, Reynolds' equation (Eq. 7.9) will be used. Integration of this with respect to r, under the boundary condition that the pressure gradient at the disk center is zero, i.e.,

$$\frac{\partial p}{\partial r} = 0 \quad \text{at} \quad r = 0,$$

yields the following equation:

$$\frac{\partial p}{\partial r} = \frac{6\mu r \dot{h}}{h^3} \tag{7.17}$$

Another integration of this with respect to r under the boundary condition:

$$p = 0 \quad \text{at} \quad r = r_a$$

gives the fluid pressure as follows:

$$p = \frac{3\mu \dot{h}}{h^3}(r^2 - r_a^2) \tag{7.18}$$

In other words, when the fluid inertia can be neglected, the fluid film pressure is proportional to the coefficient of viscosity and the approaching velocity of the two surfaces, and is inversely proportional to the third power of the film thickness. Further, the pressure distribution in the radial direction will be a parabola which has the maximum at the center of the disk. The pressure p is positive when $\dot{h}$ is negative (positive squeeze).

Integration of Eq. 7.18 over the disk gives the load capacity P as follows:

$$P = \int_0^{r_a} 2\pi r p \, dr = -\frac{3\pi}{2} \frac{\mu \dot{h} r_a^4}{h^3} \tag{7.19}$$

Now, let us consider the fluid velocity. The fluid velocity v_r in the radial direction can be obtained from Eqs. 7.8 and 7.17 as follows:

$$v_r = \frac{3r\dot{h}}{h^3}(z^2 - hz) \tag{7.20}$$

That is, v_r obeys a parabolic distribution in the thickness direction and is highest at the middle of the film thickness. The fluid velocity v_z in the thickness direction can be found from the continuity equation Eq. 7.4 and Eq. 7.20 under the boundary condition $v_z = 0$ at $z = 0$, $v_z = \dot{h}$ at $z = h$ as follows:

$$v_z = -\frac{\dot{h}}{h^3}(2z^3 - 3hz^2) \tag{7.21}$$

These are the basic equations for a squeeze film when the fluid inertia is neglected.

7.2.2 Squeeze with Fluid Inertia

When the fluid inertia is not negligible, the Navier–Stokes equation must be solved and, as stated before, three equations, Eqs. 7.3, 7.4, and 7.5 (or Eq. 7.6), will be the basic equations for the problem.

The pressure in this case can be obtained by adding modifying terms due to the fluid inertia to the solution in the previous section where fluid inertia was neglected. First obtain $\partial v_r/\partial t$, $\partial v_r/\partial r$, and $\partial v_r/\partial z$ from the equations of fluid velocity, Eqs. 7.20 and 7.21, then substitute $\partial v_r/\partial t$, $\partial v_r/\partial r$, and $\partial v_r/\partial z$ into Eq. 7.3 and integrate it twice with respect to z assuming that $\partial p/\partial r$ does not depend on z, then v_r, which includes $\partial p/\partial r$, will be obtained. Substituting the result into Eq. 7.5 and integrating once again, we obtain the first modification of the pressure distribution taking inertia into consideration as follows:

$$p = \left(\frac{3\mu \dot{h}}{h^3} + \frac{3\rho \ddot{h}}{10h} - \frac{15\rho \dot{h}^2}{28h^2}\right)(r^2 - r_a^2) - \Delta p \tag{7.22}$$

where $\Delta p = 0$ in the case of positive squeeze. The first term in the parenthesis of the right-hand side of the above equation is a viscous solution, and the second and the third terms are modifications arising from inertia.

The second modification of the pressure can be obtained by repetition of a similar procedure using velocities v_r and v_z calculated from the first modification, Eq. 7.22. The calculations are, however, very troublesome.

If the squeeze Reynolds number Re_s becomes very large, viscosity can be neglected compared with inertia. The definition of Re_s is as follows with $V = -\dot{h}$:

$$Re_s = hV/\nu \tag{7.23}$$

Then, Eq. 7.3 will be as follows:

$$\frac{\partial p}{\partial r} = -\rho\left(\frac{\partial v_r}{\partial t} + v_r\frac{\partial v_r}{\partial r} + v_z\frac{\partial v_r}{\partial t}\right) \tag{7.24}$$

In the case of an ideal fluid, since $v_r = -r\dot{h}/(2h)$, the above equation gives the pressure considering only the inertia of the fluid as:

$$p = \left(\frac{\rho\ddot{h}}{4h} - \frac{3\rho\dot{h}^2}{8h^2}\right)(r^2 - r_a^2) \tag{7.25}$$

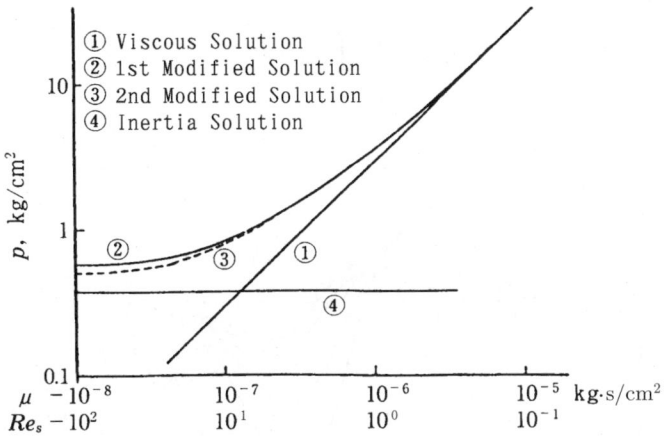

Fig. 7.3. Comparison of viscous, modified, and inertia solutions [7]

Figure 7.3 compares Eqs. 7.18, 7.22, and 7.25 in a constant velocity squeeze in which $r_a = 10$ cm, $h = 0.1$ cm, $\dot{h} = -10$ cm/s, $\rho = 10^{-6}$ kg·s^2/cm^4, and $\mu = 10^{-8} - 10^{-4}$ kg·s/cm^2. Both μ and the squeeze Reynolds number $Re_s = hV/\nu$ are taken on the horizontal axis, and the pressure at the center of the squeeze surface p is taken on the vertical axis. The figure shows that the modified solution which takes inertia into consideration is close to the viscous solution when the Reynolds number is small and close to the inertia solution when the Reynolds number is large. The figure also shows the second modified solution (calculated also for a positive uniform squeeze), which is not very different from the first modified solution in the range shown in the figure.

7.2.3 Sinusoidal Squeeze Motion

Let us consider a squeeze film in which the gap between two surfaces or the film thickness changes sinusoidally. In this case, the film thickness h will be given as:

$$h = h_0 + h_a(\cos 2\pi f t - 1) \tag{7.26}$$

where h_0 is an initial film thickness (the maximum film thickness), h_a is the amplitude of the sinusoidal change of the film thickness, and f is its frequency. Further, define an average Reynolds number Re_o for the sinusoidal squeeze as follows, ν being the coefficient of kinetic viscosity:

$$Re_o = h_0^2 f/\nu \tag{7.27}$$

The intensity of a sinusoidal squeeze depends on Re_o and h_a/h_0.

Substitution of Eq. 7.26 into Eq. 7.18 when fluid inertia is neglected, or into Eq. 7.22 when fluid inertia is taken into account, yields pressure p in the case of a sinusoidal squeeze. The integration of p over the squeeze surface gives the load capacity P as:

$$P = 2\pi \int_0^{r_a} p\, r dr \tag{7.28}$$

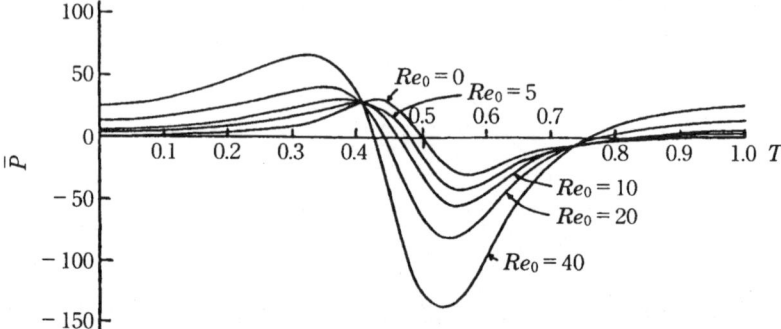

Fig. 7.4. Variation of nondimensional load capacity over a cycle [7]

Figure 7.4 shows the variation of the nondimensional load capacity $\bar{P} = P/(12\mu r_a^4 f/h_0^2)$ in one cycle of a sinusoidal squeeze, where $h_a/h_0 = 0.4$ is assumed. The horizontal axis shows the nondimensional time $T = ft$. The parameter Re_o in the figure is the average Reynolds number for a sinusoidal squeeze. In the case of $Re_o = 0$, only viscosity is at work, and the curve of $\bar{P}$ is one of point symmetry with respect to nondimensional time $T = 0.5$, as shown in the figure. In this case, the squeeze speed is zero at nondimensional time $T = 0, 0.5$, and 1.0, and since fluid inertia is neglected, the value of $\bar{P}$ is also zero. As Re_o becomes large, however, under the influence of fluid inertia, the curve of nondimensional load capacity changes shape considerably. In this case, while the constituent of $\overline{P}$ at $T = 0$, 0.5, and 1.0 attributable to viscosity is zero, that attributable to inertia increases greatly (at these time points, the acceleration of a squeeze surface is maximum).

7.3 Sinusoidal Squeeze by a Rigid Surface (Experiments)

Analysis of a squeeze film is easy in the case of positive squeeze; however, in the case of negative squeeze, cavitation may occur in the fluid film and the analysis becomes difficult. In this section, a sinusoidal squeeze is investigated experimentally (Kuroda and Hori [9]).

7.3.1 Mild Sinusoidal Squeeze

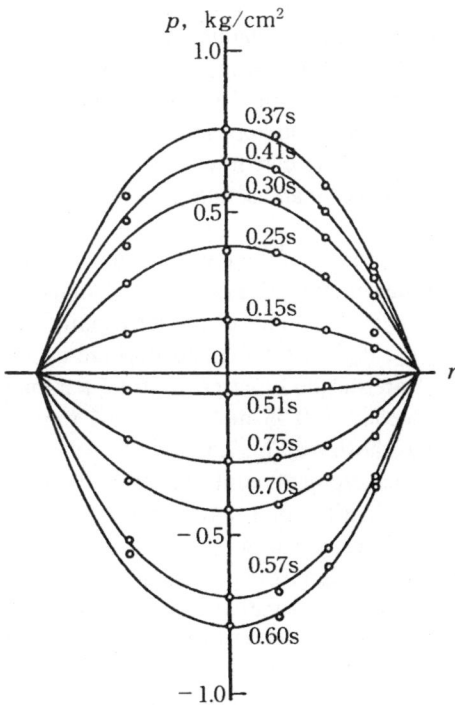

Fig. 7.5. Variation of pressure distribution in one cycle [9]

Figure 7.5 shows an example of experimentally obtained variation of pressure distribution in one cycle of mild sinusoidal squeeze for disks. The squeeze surface

146   7 Squeeze Film

is a disk of diameter $2r_a = 116$ mm, the initial film thickness (maximum film thickness) is $h_0 = 1.0$ mm, the amplitude of the sinusoidal squeeze is $h_a = 0.3$ mm, the frequency is $f = 1$ Hz, and the lubricant used is SAE No. 90. As the theory shows, the pressure distribution is parabolic and is symmetric about the horizontal axis. Following the parameter, which shows time in seconds, we can see that the pressure rises from zero first, reaches a positive peak value, then lowers and reaches a negative peak value, and returns to zero again after 1 s. Positive and negative peak values of the pressure are 0.76 kg/cm² and −0.80 kg/cm², respectively, and coincide approximately with the theoretical value of 0.80 kg/cm².

7.3.2 Intense Sinusoidal Squeeze — Cavitation

As the sinusoidal squeeze becomes more intense, the pressure generated becomes large in both positive and negative senses. For positive squeeze, there is no particular upper limit to the pressure generated, but for negative squeeze, cavitation may occur in the fluid film when the pressure becomes lower than a certain limit, and the pressure will not fall any further. A similar phenomenon will occur when gas molecules dissolved in the fluid separate and form air bubbles. For this reason, even if the squeeze motion is symmetric in the positive and negative directions, the positive–negative symmetry of generated pressure will be lost. For the stationary state, the pressure at which cavitation appears is the vapor pressure of the fluid, but in a dynamic situation, the pressure may transiently go significantly below the vapor pressure, and may even be lower than vacuum pressure. In the latter case, tension is generated.

Examples of the time variation of the pressures in such a case are shown in Fig. 7.6. The pressures were measured at 12 points on the squeeze surface shown in the attached figure. Point P_1 is at the center of the disk, points (P_3, P_6, and P_9) are on the inner circle, points (P_2, P_4, P_5, P_7, P_8, and P_{10}) are on the intermediate circle, and points (P_{11} and P_{12}) are on the outermost circle. Parameters of the sinusoidal movement in this case are $h_0 = 0.95$ mm, $h_a = 0.4$ mm, $f = 2$ Hz, and other parameters are the same as those for Fig. 7.5.

The pressure at each measurement point changes smoothly with time during positive squeeze. The pressure distribution is axisymmetric and parabolic with the maximum at the center of the squeeze surface. This is as predicted by theories. When the squeeze motion changes its sign from positive to negative, the pressure generated will also change from positive to negative. The pressure lowers gradually, passes zero pressure, and at $t = 0.290$ s (the point of sharp downward projection), tension (pressure below −1 atmospheric pressure) appears at the all points of measurement. Cavitation occurs when the oil film cannot bear the tension any more, and the pressure returns rapidly to a constant value near the vapor pressure of the fluid. The pressure returns to atmospheric pressure gradually after that. If the magnitude of the tension, the time of appearance and the duration of the tension are checked carefully, it turns out that they differ at the each point of measurement and hence the pressure distribution becomes nonaxisymmetric for negative squeeze. In this particular exam-

7.3 Sinusoidal Squeeze by a Rigid Surface (Experiments) 147

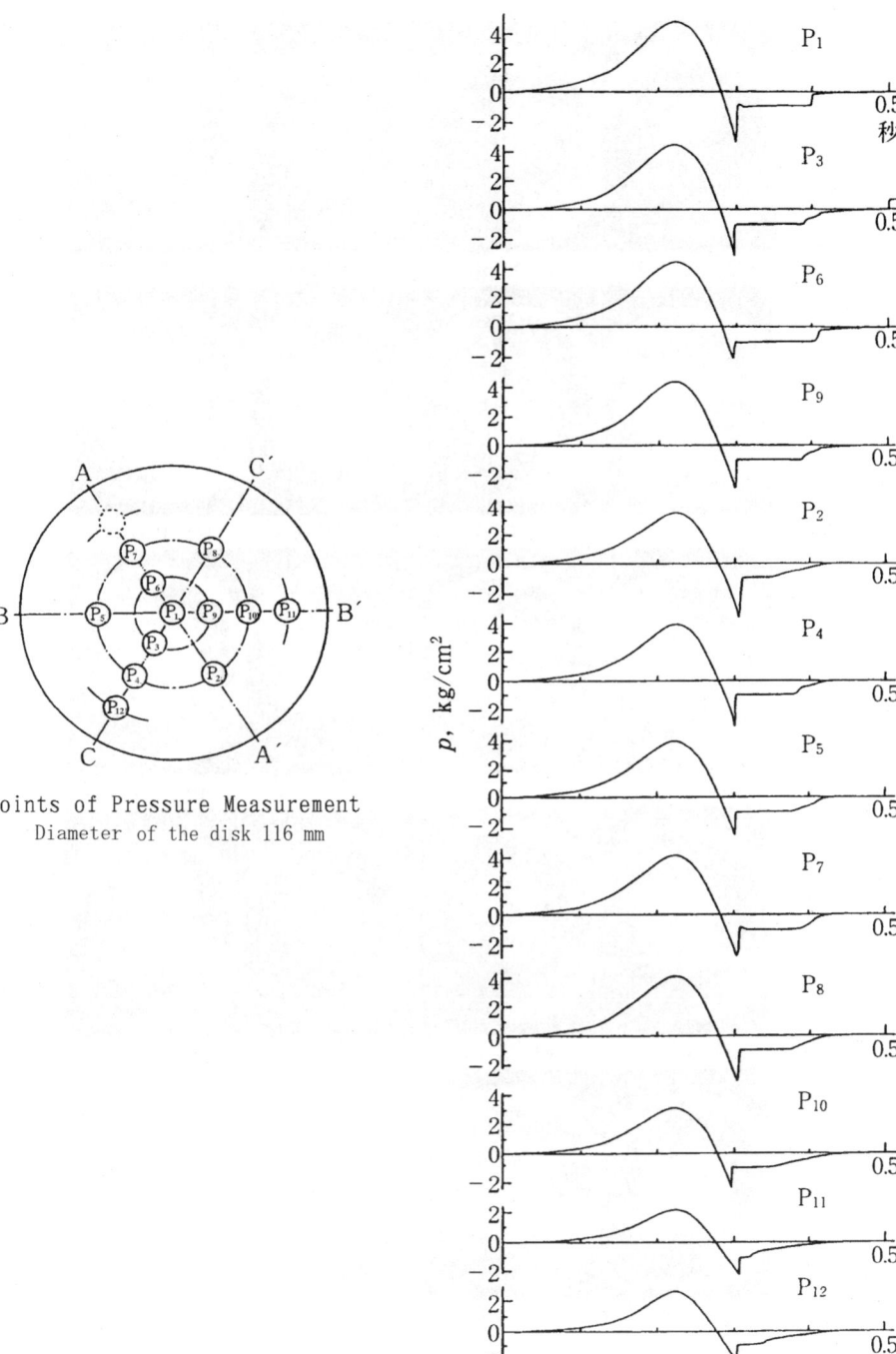

Points of Pressure Measurement
Diameter of the disk 116 mm

Fig. 7.6. Pressure change at measuring points for an intense sinusoidal squeeze [9]

148 7 Squeeze Film

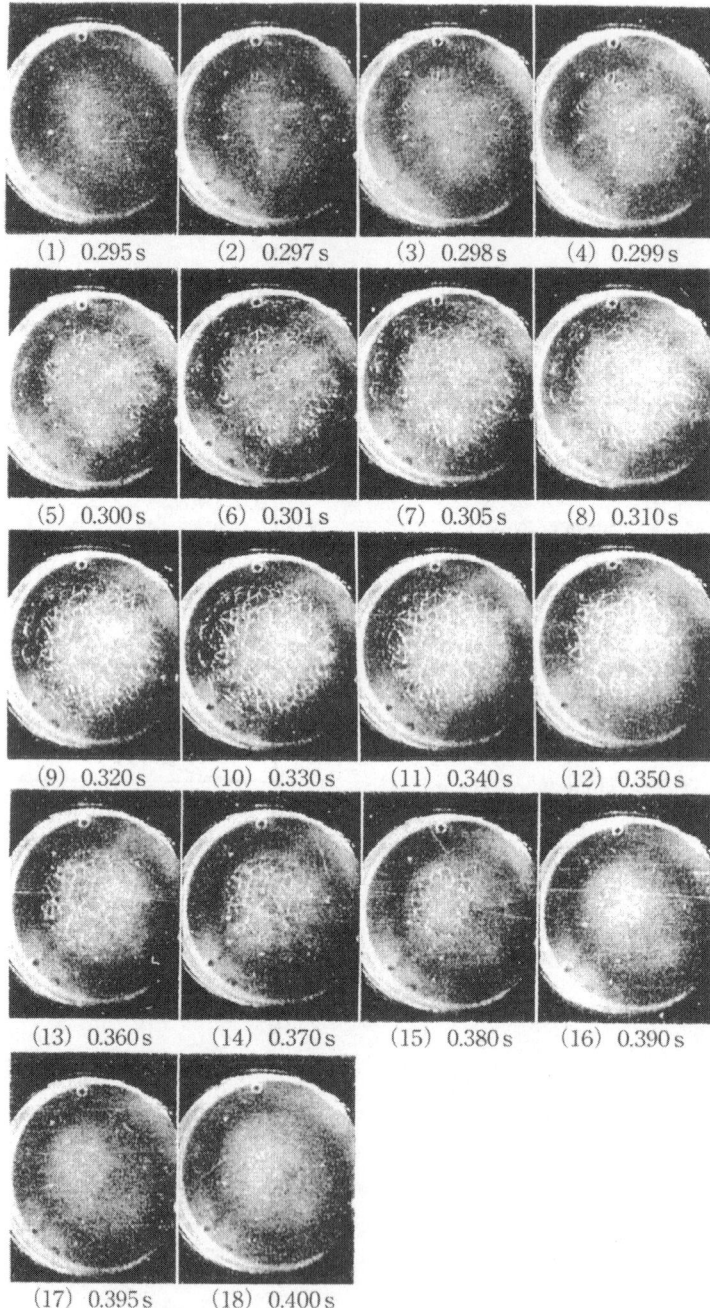

Fig. 7.7. Serial photographs of cavitation [9]

ple, the lowest negative pressure is -3.5 kg/cm^2 and is generated at P$_2$ (to the lower right of the point P$_1$, i.e., the fifth curve from the top).

Figure 7.7 is a series of photographs showing the situation in the oil film from the appearance to the disappearance of the cavitation taken with a high-speed camera through the squeeze surface (glass plate) simultaneously with the pressure measurement in Fig. 7.6. The first air bubbles appear in frame (1), $t = 0.295$ s, a little after the squeeze changes from positive to negative. Cavitation is not the result of growth of a single bubble, but of many bubbles that originate at various points and quickly grow and, in the 0.015 s before frame (8) at $t = 0.310$ s, cover almost the whole area inside a circle of a radius of three-quarters that of the squeeze surface. Cavitation appears and grows quickly in this way, and then disappears slowly. In the latter process, air bubbles shrink gradually with the recovery of pressure.

Although the pressure recovers to atmospheric pressure at the end of one cycle of squeeze motion, some of the air bubbles generated during cavitation remain inside the squeeze surface. This may be attributed to the separation of gas molecules from the fluid at low pressure.

7.4 Sinusoidal Squeeze with a Soft Surface

The squeeze of a fluid film between the bottom surface of a rubber block and a rigid surface as shown in Fig. 7.8 is considered next (Nakano et al. [6], and Hori, Kato et al. [10] [12]).

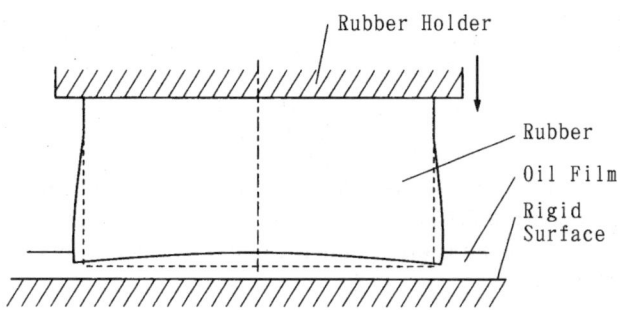

Fig. 7.8. Squeeze with a soft surface [12]

The rigid surface is stationary and a sinusoidal motion is applied to the rubber block through a holder. During positive squeeze, the squeeze surface of the rubber block becomes concave due to the pressure produced in the fluid film, as shown in the figure, so that the fluid cannot flow out easily. During negative squeeze, in constrast to this, the squeeze surface becomes convex and hence the fluid can flow in easily. For this reason, even if the motion given to the holder is sinusoidal and positive–negative symmetric, the time average of the pressure generated in the fluid film is positive, so that a positive load capability is produced. Let us consider this phenomenon in more detail below. It should be noted that, while rubber can be treated as an elastic body when the frequency of sinusoidal motion is low, it must be treated as a viscoelastic body when the frequency is high.

7.4.1 Low-Frequency Squeeze

In calculating the deformation of a rubber block as shown in Fig. 7.8, consider a low frequency squeeze and assume rubber be an elastic body (not viscoelastic).

The stress–strain relation of an elastic body is usually written as [3]:

$$\sigma_{ij} = \frac{2G\nu}{(1-2\nu)} \epsilon_v \delta_{ij} + 2G \epsilon_{ij} \tag{7.29}$$

where ν is Poisson's ratio here, ϵ_v is the volumetric strain, G is the shear modulus, and δ_{ij} is Kronecker's delta: $\delta_{ij} = 1$ if $i = j$ and $\delta_{ij} = 0$ if $i \neq j$. Poisson's ratio ν for rubber is very close to 0.5. This means that rubber deforms with almost no volume change, because (volumetric modulus)$\gg$ (shear modulus) for rubber. If Poisson's ratio ν nears 0.5 in the above equation, the first term of the right-hand side will be an indeterminate form of $(\infty \times 0)$, and so the error in numerical computation may become very large. To avoid this difficulty, Hermann's method will be used [2] [4]. This method is applicable to any Poisson's ratio between 0 and 0.5, including 0.5. However, an ordinary elasticity calculation can also give a reasonable approximation, except for some special cases, if a Poisson's ratio is used that is close to 0.5 but can still guarantee stable numerical calculations (e.g., 0.495) [8] [11].

In Herrmann's method, the following mean stress function H is introduced:

$$H = \frac{3\sigma_m}{2G(1+\nu)} \tag{7.30}$$

where $\sigma_m = (\sigma_x + \sigma_y + \sigma_z)/3$ is called the mean stress. Substituting the relation between elastic coefficients $G = [3K(1-2\nu)]/[2(1+\nu)]$ (where K is the volumetric modulus) into Eq. 7.30 yields the volumetric strain $\epsilon_v = \sigma_m/K$ as:

$$\epsilon_v = (1-2\nu)H \tag{7.31}$$

Substituting this into Eq. 7.29 gives the following equation, $(1-2\nu)$ being eliminated:

$$\sigma_{ij} = 2G(\epsilon_{ij} + \nu H \delta_{ij}) \tag{7.32}$$

The above equation can be used for all Poisson's ratios of $0 < \nu \leq 0.5$.

With body force F^i, displacement u_i, and surface force P^i, the following functional Π, which includes the mean stress function H, is introduced here:

$$\Pi = \int_v \left[G\{I^2 - 2II + 2\nu HI - \nu(1-2\nu)H^2\} - F^i u_i \right] dV - \int_s P^i u_i dS \tag{7.33}$$

where I and II are the strain invariants of first order and second order, respectively. The solutions (displacement u_i, mean stress function H) for an elasticity problem can then be obtained by solving:

$$\delta \Pi = 0 \tag{7.34}$$

where $\delta \Pi$ is a variation of the functional Π.

In rectangular coordinates (x, y, z), the strain invariant of Eq. 7.33 can be written as:

$$I = \epsilon_x + \epsilon_y + \epsilon_z \tag{7.35}$$

$$I^2 - 2II = (\epsilon_x^2 + \epsilon_y^2 + \epsilon_z^2) + 2(\epsilon_{xy}^2 + \epsilon_{yz}^2 + \epsilon_{zx}^2) \tag{7.36}$$

Further, in cylindrical coordinates (r, θ, z), the strain invariant can be written as:

$$I = \epsilon_r + \epsilon_\theta + \epsilon_z \tag{7.37}$$

$$I^2 - 2II = (\epsilon_r^2 + \epsilon_\theta^2 + \epsilon_z^2) + 2(\epsilon_{zr}^2 + \epsilon_{r\theta}^2 + \epsilon_{\theta z}^2) \tag{7.38}$$

where the strains are given as follows in physical components [1] [3]:

$$\epsilon_r = \frac{\partial u_r}{\partial r}, \quad \epsilon_\theta = \frac{u_r}{r} + \frac{1}{r}\frac{\partial u_\theta}{\partial \theta}, \quad \epsilon_z = \frac{\partial u_z}{\partial z}, \quad \epsilon_{zr} = \frac{1}{2}\left(\frac{\partial u_z}{\partial r} + \frac{\partial u_r}{\partial z}\right),$$

$$\epsilon_{r\theta} = \cdots, \quad \epsilon_{\theta z} = \cdots$$

An axisymmetrical cylindrical rubber block is considered here in cylindrical coordinates, and is divided into ring elements of a triangular cross section as shown in Fig. 7.9. In an axisymmetrical case, $(1/r)(\partial u_\theta/\partial \theta) = 0$, $\epsilon_{r\theta} = 0$, and $\epsilon_{\theta z} = 0$ hold in the above strain equations.

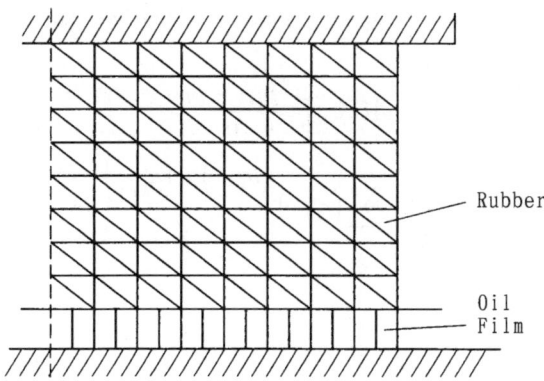

Fig. 7.9. Mesh division of a cylindrical rubber block [12]

Discretizing the functional in Fig 7.33 and applying the finite element method give the following relation at each nodal point (body forces are disregarded):

7 Squeeze Film

$$f_i = a_{ij}d_j + b_{ij}H_j \tag{7.39}$$

where d_j and f_j are the displacement and the force applied at the jth nodal point and H_j is the mean stress function of the jth element.

In the present case, the only external force considered is that of the fluid film pressure and the only displacement of interest is the vertical displacement of the bottom of the rubber cylinder. Therefore, it is convenient to contract Eq. 7.39 in the following form:

$$p_i = S_{ij}\delta_j \tag{7.40}$$

where δ_j and p_i are the displacement and pressure at the nodal points of the bottom of the cylinder, respectively.

Now, Reynolds' equation can be written in cylindrical coordinates as:

$$\frac{\partial}{\partial r}\left(rh^3\frac{\partial p}{\partial r}\right) = 12\mu r\dot{h} \tag{7.41}$$

Giving a sinusoidal motion to the above rubber holder, we can write the fluid film thickness as:

$$h = h_0 + h_a \cdot \sin(2\pi ft) + \delta \tag{7.42}$$

where h_0 is the initial film thickness (average film thickness), h_a is the amplitude of the sinusoidal motion of the rubber holder, and δ is the deflection of the bottom surface of the rubber.

Integration of Reynolds' equation (Eq. 7.41) with respect to r results in:

$$rh^3\frac{\partial p}{\partial r} = 12\mu \int_0^r r\dot{h}dr \tag{7.43}$$

Let us divide the lubricating film into a series of rings as shown in Fig. 7.9 and also divide time into a series of time steps Δt. Then we can discretize the above equation as follows.

$$\left(r_{i+\frac{1}{2}}\right)\left(h_{i+\frac{1}{2}}\right)^3 \frac{p_{i+1} - p_i}{\Delta r_i}$$
$$= 12\mu \sum_{k=1}^{i} \left(r_{k+\frac{1}{2}}\right) \frac{\left(h_{k+\frac{1}{2}}\right) - \left(h'_{k+\frac{1}{2}}\right)}{\Delta t}\Delta r_i \tag{7.44}$$

where

$$h_{i+\frac{1}{2}} = \frac{1}{2}(h_i(t) + h_{i+1}(t))$$
$$h'_{i+\frac{1}{2}} = \frac{1}{2}\left(h_i(t - \Delta t) + h_{i+1}(t - \Delta t)\right) \tag{7.45}$$
$$r_{i+\frac{1}{2}} = \frac{1}{2}(r_i + r_{i+1})$$
$$\Delta r_i = r_{i+1} - r_i$$
$$(i = 1, 2, \ldots, N)$$

7.4 Sinusoidal Squeeze with a Soft Surface   153

We then solve Eqs. 7.40 and 7.44 simultaneously under the boundary condition

$$p_{N+1} = 0 \tag{7.46}$$

using the Newton–Raphson method.

Since Eq. 7.44 includes h'_i, the time change of the pressure distribution and that of the fluid film shape cannot be obtained by a single iterative calculation. The equation is calculated from the beginning for every time step, and the h_i value obtained is used as h'_i for the next time step.

7.4.2 High-Frequency Squeeze

When the frequency of the sinusoidal squeeze motion becomes high, the viscoelasticity of the rubber cannot be ignored. For high frequencies, the apparent elastic coefficients increase and the phase difference between stress and strain becomes significant.

Consider a square rubber block. For simplicity, divide the block into many columns (pillars) as shown in Fig. 7.10, and assume that each column deforms in the axial direction only and independently from each other. Also, assume that the dynamic characteristics of rubber can be expressed by the spring–dashpot models of three elements, four elements, and five elements shown in Fig. 7.10, and that the dynamic behavior of a column can be expressed by the following constitutive equation:

$$\sigma + a_1 \dot{\sigma} + a_2 \ddot{\sigma} = E(\epsilon + b_1 \dot{\epsilon} + b_2 \ddot{\epsilon}) \tag{7.47}$$

where E is Young's modulus.

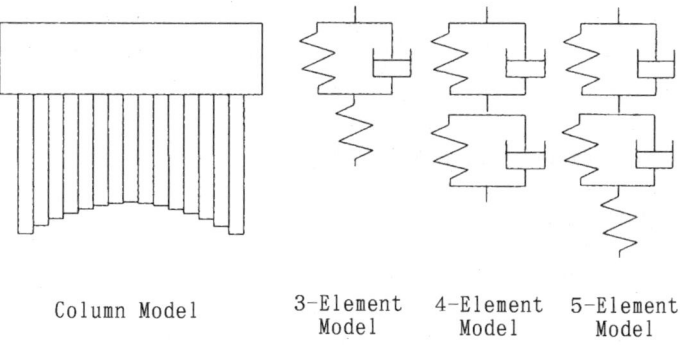

Column Model   3-Element   4-Element   5-Element
                 Model   Model   Model

Fig. 7.10. Column model and viscoelastic models of rubber

The stress required to deform a column at a constant strain rate can be expressed by the following function of time:

$$\sigma = c_1 + c_2 t + c_3 \exp(-t/r_1) + c_4 \exp(-t/r_2) \tag{7.48}$$
$$= F(t) \tag{7.49}$$

$F(t)$ in the above equation is called the constant-strain-rate modulus. The coefficients c_i ($i = 1, 2, 3, 4$) and r_i ($i = 1, 2$) of the above equation can be expressed in terms of the coefficients a_1, a_2, b_1, b_2 of Eq. 7.47 and E. Using F, the stress σ_n after an arbitrary strain history can be expressed approximately as:

$$\sigma_n = \sum_{i=1}^{n} F(t_n - t_{i-1}) \frac{\epsilon_i - 2\epsilon_{i-1} + \epsilon_{i-2}}{\Delta t} \tag{7.50}$$

where σ_i and ϵ_i are the stress and strain, respectively, at time $t = t_i = i \cdot \Delta t$. The strain before $t = t_0$ has been assumed to be zero. Equation 7.50 states that stress σ_n at an arbitrary time t_n can be expressed in terms of all the previous strains, i.e., the strains at times $t_j \le t_n$.

Meanwhile, Reynolds' equation for a square squeeze surface is:

$$\frac{\partial}{\partial x}\left(\frac{h^3}{\mu}\frac{\partial p}{\partial x}\right) + \frac{\partial}{\partial y}\left(\frac{h^3}{\mu}\frac{\partial p}{\partial y}\right) = 12\dot{h} \tag{7.51}$$

where

$$h = h_0 + h_a \cos(2\pi f t) + \delta \tag{7.52}$$

in which h_0 is the average film thickness, h_a is the amplitude of sinusoidal motion of the rubber holder, and δ is the deflection of the bottom surface of the rubber.

The functional of Eq. 7.51 is discretized by the finite element technique and the Ritz procedure is applied to it [5]. We then combine the result with Eq. 7.50 (σ_n is replaced by p_n) and solve it numerically by the Newton–Raphson iteration method. As boundary conditions, it is assumed that $p = 0$ at the periphery of the bottom of the rubber block. Numerical computation is performed at each time step in the same way as in the previous section.

7.4.3 Results of Experiment and Calculation

a. Low-Frequency Squeeze

An experiment using a low frequency sinusoidal squeeze was carried out with a cylindrical rubber block 116 mm in diameter and 50 mm high. The pressure experimentally obtained and that calculated by the theory of Section 7.4.1 are compared in Figs. 7.11 and 7.12. Young's modulus and Poisson's ratio for the rubber are $E = 0.8$ MPa and $\nu = 0.5$, respectively, and the coefficient of viscosity of the fluid is $\mu = 320$ cP.

Figure 7.11 shows the time variation of the pressure at the center of the bottom of the rubber block in the first and sixth cycle of a sinusoidal squeeze. The parameters of squeeze motion were as follows: initial thickness of the sinusoidal motion $h_0 =$

7.4 Sinusoidal Squeeze with a Soft Surface   155

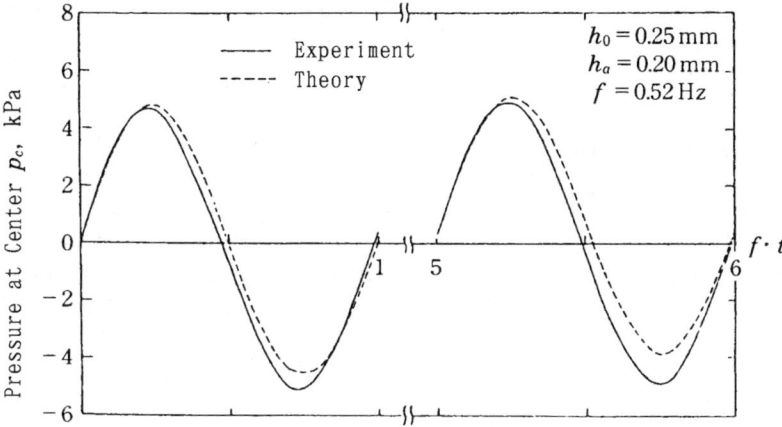

Fig. 7.11. Time variation of the pressure in the 1st and 6th cycle [12]

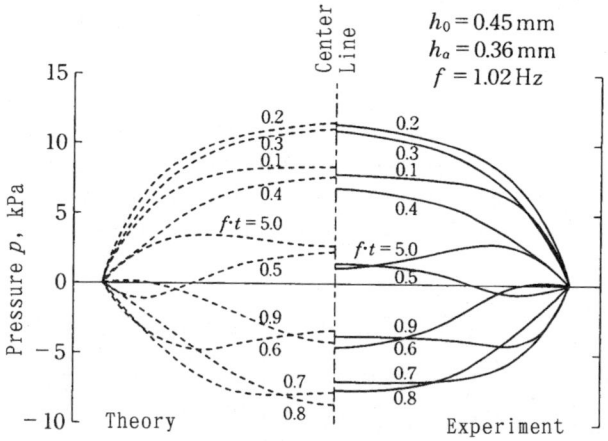

Fig. 7.12. Time variation of the pressure distribution in the 6th cycle [12]. *solid lines*, experimental results; *dashed lines*, theoretical calculations

0.25 mm, amplitude of the sinusoidal motion $h_a = 0.20$ mm, frequency $f = 0.52$ Hz. Solid lines show experimental results and dashed lines show theoretical calculations.

Figure 7.12 shows the time variation of the pressure distribution during the fluid film during the sixth cycle. The parameters of squeeze motion were as follows: initial thickness of the fluid film $h_0 = 0.45$ mm, amplitude of the sinusoidal motion $h_a = 0.36$ mm, frequency $f = 1.02$ Hz. Solid lines show experimented results (the right half), and dashed lines show theoretical calculations (the left half).

In both figures, experiment and the theory based on the assumption that the rubber is elastic are in good agreement. This shows that rubber can be treated as elastic at this frequency.

b. High-Frequency Squeeze

An experiment using a high-frequency sinusoidal squeeze was carried out with a square rubber block 120 mm × 120 mm × 20 mm. Experimental results and the theory of Section 7.4.2 are compared in Figs. 7.13 and 7.14. The parameters of the squeeze motion are as follows: $h_0 = 0.18$ mm, $h_a = 0.12$ mm, $f = 18.2$ Hz, coefficient of viscosity $\mu = 130$ cP.

Table 7.1. Coefficients of the constitutive equation of rubber

Model	E (kgf/cm^2)	a_1 (s)	a_2 (s^2)	b_1 (s)	b_2 (s)
three-element	80.5	1.68×10^{-2}	0.0	2.40×10^{-2}	0.0
four-element	80.5	2.63×10^{-2}	0.0	3.63×10^{-2}	6.94×10^{-5}
five-element	85.1	3.46×10^{-2}	6.22×10^{-5}	4.19×10^{-2}	1.30×10^{-4}

Coefficients of the constitutive equation of the rubber (Eq. 7.47) are given in Table 7.1. These values were experimentally determined by applying oscillatory compression (frequency range 0.01 – 38 Hz) to the rubber and approximating the stress response by three-element, four-element, and five-element models.

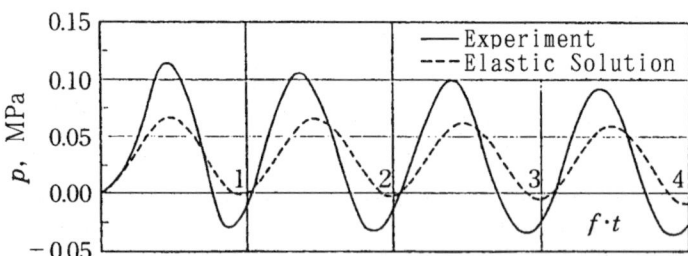

Fig. 7.13. Time variation of the pressure when rubber is assumed to be elastic [10] [12]

The time variation of the calculated pressure at the center of the bottom of the rubber block is compared with experimental results in Fig. 7.13. In the calculation, only the elasticity of the rubber was considered (a_1, a_2, b_1 and b_2 in Table 7.1 are assumed to be zero). The highest pressure in the experiment (solid lines) is 1.5 – 1.7 times higher (i.e., the rubber is harder) than that in the calculation (dashed lines), and the highest pressure appears earlier in the experiment than in the calculation.

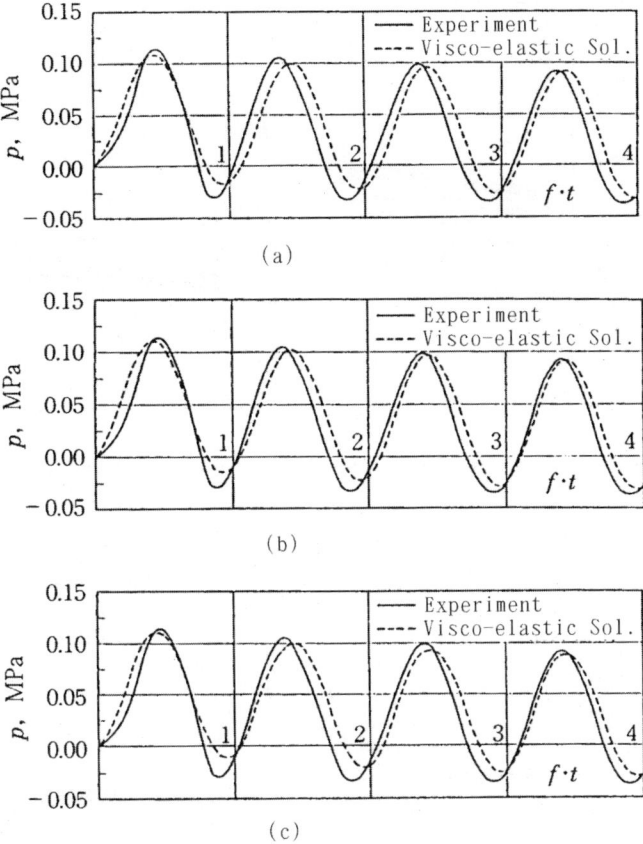

Fig. 7.14a–c. Time variation of the pressure when the rubber is assumed to be viscoelastic [10] [12]. **a** three-element model, **b** four-element model, **c** five-elemnt model

We next consider the rubber to be viscoelastic and carry out similar calculations using the three kinds of viscoelastic model. Figure 7.14 shows the comparisons of the experimental results and the calculations. They are in good agreement this time for each model. These figures show that, in this case, the three-element model is adequate.

It is seen in the figures that the time average of the pressure is not zero but is greatly shifted upward. It is interesting to note that although the squeeze motion is positive–negative symmetric, a large load capability is obtained.

158   7 Squeeze Film

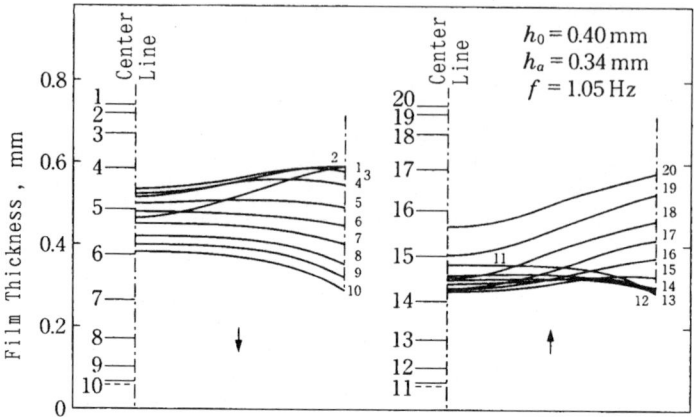

Fig. 7.15. Time variation of the shape of the bottom surface of a rubber block (f = 1.05 Hz) [12]

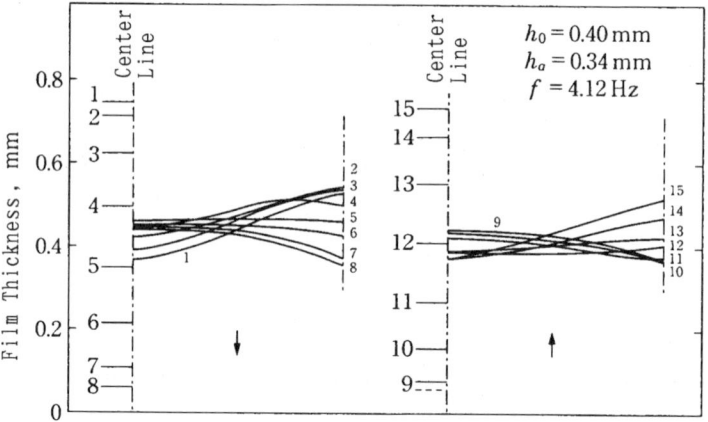

Fig. 7.16. Time variation of the shape of the bottom surface of a rubber block (f = 4.12 Hz) [12]

c. Deformation of the Bottom Surface of the Rubber

Examples are shown of the measured time variation in the shape of the bottom surface of the rubber for the low-frequency squeeze analyzed in paragraph **a.** of this section.

The moiré method (see Section 6.4.2) is used for the measurement of the deformation of the bottom surface of the rubber block. In this connection, squeeze of an oil film between the bottom surface of the rubber and a glass plate (assumed to be rigid) with a grating of a line density of 400 lines/inch is considered. The fringes obtained in this case are concentric circles corresponding to contour lines of the rubber surface and the difference of heights (spacing between the glass plate and the rubber surface) between two adjacent fringe lines is about 63 μ. When a sinusoidal motion is given to the rubber block, the concentric circles repeat centripetal and centrifugal movements, according to the motion of the rubber surface. The situation was recorded with a video and the change of the bottom shape, or that of the oil film, was analyzed.

Figures 7.15 and 7.16 show the time variation of the oil film thickness distribution obtained from analysis of the moiré pattern for frequencies $f = 1.05$ Hz and 4.12 Hz, respectively. The left half and the right half of each figure show the film thickness during positive squeeze (downward) and negative squeeze (upward), respectively. The scale on the center line shows the position of the bottom surface assuming that the rubber does not deform, and the numbers accompanying the scale correspond to those accompanying the curves of the oil film shape. The experimental conditions not shown in the figure are the same as those in paragraph **a.** of this section.

It is seen in the figure that the bottom surface of the rubber is concave during positive squeeze and convex during negetive squeeze, and altogether it flutters like a bird's wings. As a result, the time average of the fluid pressure over several cycles of squeeze becomes positive and a considerable load capacity arises. It is seen that the bottom surface is convex in the early stages of a positive squeeze in both figures. This is a carry over of the deformation of the bottom surface from the previous cycles. Further, the comparison of the two figures shows that the amplitude of the movement of the bottom surface (variation of film thickness) is smaller when the frequency is high, particularly at the center of the bottom surface.

References

1. Y. Yamamoto, "Elasticity and Plasticity" (in Japanese), Asakura Shoten, 1961, Tokyo.
2. L.R. Herrmann and R.M. Toms, "A Reformulation of the Elastic Field Equation, in Terms of Displacements, Valid for all Admissible Values of Poisson's Ratio", *Trans. ASME, Journal of Applied Mechanics*, March 1964, Vol. 31, pp. 140 - 141.
3. Y.C. Fung, "Foundations of Solid Mechanics", Prentice-Hall, Inc., Englewood Cliffs, N.J., 1965.
4. L.R. Herrmann, "Elasticity Equations for Incompressible and Nearly Incompressible Materials by Variational Theorem", *AIAA Journal*, Vol. 3, No. 10, October 1965, pp. 1896 - 1900.
5. M.M. Reddi, "Finite-Element Solution of the Incompressible Lubrication Problem", *Trans. ASME, Journal of Lubrication Technology*, Vol. 91, July 1969, pp. 524 - 533.
6. E. Nakano and Y. Hori, "Squeeze Film: The Effect of the Elastic Deformation of Parallel Squeeze Film Surfaces", *Proc. of the JSLE-ASLE International Lubrication Conference*, Tokyo June 9 - 11, 1975, pp. 325 - 332.
7. S. Kuroda and Y. Hori, "A Study of Fluid Inertia Effects in a Squeeze Film" (in Japanese), *Journal of Japan Society of Lubrication Engineers*, Vol. 21, No. 11, November 1976, pp. 740 - 747.

8. S. Kuroda, "A Study on Squeeze Film Effects (Effect of Elastic Deformation of Squeeze Surface)" (in Japanese), A Paper of *53rd Annual Meeting of the Kansai Branch of JSME*, Rm. 6, March 16 - 17, 1978, Kobe, pp. 70 - 72.
9. S. Kuroda and Y. Hori, "An Experimental Study on Cavitaition and Tensile Stress in a Squeeze Film" (in Japanese), *Journal of Japan Society of Lubrication Engineers*, Vol. 23, No. 6, June 1978, pp. 436 - 442.
10. Y. Hori and T. Kato, "A Study on Visco-Elastic Squeeze Films" (in Japanese), *Journal of Japan Society of Lubrication Engineers*, Vol. 24, No. 3, March 1979, pp. 174 - 181.
11. H. Narumiya and Y. Hori, "Deformation Analysis of An Incompressible Elastic Body by FEM" (in Japanese), A Paper of *54th Annual Meeting of the Kansai Branch of JSME*, Rm. 2, March 16 - 17, 1979, Suita, pp. 16 - 18.
12. Y. Hori, T. Kato and H. Narumiya, "Rubber Surface Squeeze Film", *Trans. ASME, Journal of Lubrication Technology*, Vol. 103, July 1981, pp. 398 - 405.

8
Heat Generation and Temperature Rise

Heat generation in the oil film and the accompanying temperature rise are the most important factors in bearings. For example, temperature rise is the factor indicating the operating conditions of a bearing most directly, and if the temperature rise is small, the bearing is probably in a good operating condition. Generally speaking, the problems of heat generation and temperature rise are hard to handle, and so they were not considered in, for example, the early theory of Reynolds. It is thanks to the later development of computers that this kind of problem can now be handled theoretically.

Let us first consider the meaning of heat generation and temperature rise in bearings. To begin with, the heat generation essentially corresponds to the loss of mechanical energy due to shear in the lubricant film (solid friction is sometimes also present) of a bearing. Therefore, the less heat generated the better.

The effects of temperature rise constitute a bigger problem than the heat generation. The temperature rise decreases the viscosity of the lubricating oil, and thus the minimum film thickness and allows seizure to occur more easily. Further, the temperature rise changes the bearing clearance through the thermal deformation of the bearing metal and casing, thus changing bearing performance.

Furthermore, an even bigger problem is that the boundary lubrication performance of the lubricant film will suddenly and almost completely be lost if the oil temperature exceeds a certain critical temperature. A lubricant film has in effect a kind of transition temperature. If the oil temperature is lower than this, the molecules of lubricant combine with a metal surface strongly, and also with the adjacent molecules of lubricant, and form a strong lubricant film on the metal surface. However, if the temperature exceeds the transition temperature, these combinations are lost and the strength of lubricant film will fall markedly. Thus the performance of boundary lubrication of the oil film will be lost and seizure can take place very easily. Therefore, the oil temperature must be kept under the transition temperature, which is unfortunately relatively low (for example 100°C for low-cost oils and 160° – 170°C for high-quality oils).

In addition, if the oil temperature exceeds 150°C, the rate of oxidization (or degradation) of the lubricating oil is markedly increased. Also at 100°C, the tensile strength of white metal falls to one-half that at room temperature. Thus, it is recommended to keep the highest temperature in the bearing lower than 100° – 120°C.

In this connection, it is very important in bearing design to know accurately the highest temperature in a bearing. However, it is in fact quite difficult to achieve this, particularly in the design of new bearings. A major goal of forced lubrication in high speed or heavy load bearings is to remove the heat generated and to keep the highest temperature below the above-mentioned limit.

Let us calculate, for reference, the amount of heat generated in a journal bearing using Petrov's law (Eq. 3.32, see Chapter 3). Petrov's law assumes that the journal and the bearing are concentric. Taking a bearing for a steam turbogenerator as an example, let us consider a bearing of the following parameters: bearing diameter $D=0.60$ m, bearing length $L = 0.30$ m, mean bearing clearance $c = 0.6 \times 10^{-3}$ m (clearance ratio $c/D = 1/1000$), rotating speed $N = 3000$ rpm $= 50$ rps, and the coefficient of viscosity of the lubricating oil $\mu = 5.0 \times 10^{-2}$ Pa·s. In this case, the frictional loss or heat generated in the bearing is calculated as:

$$Q_s = \mu \left(\frac{U}{c}\right)(\pi D \cdot L)U \approx 418 \text{ kW}$$

This is a huge amount of heat. Incidentally, the circumferential speed of the journal in this case is $U= 94.2$ m/s$= 339$ km/h. It is worth noting that the surfaces of the journal and the bearing are sliding at such a large relative velocity with a separation of only 0.6 mm between them.

8.1 Basic Equations for Thermohydrodynamic Lubrication

Hydrodynamic lubrication that takes heat generation and temperature rise into consideration is called **thermohydrodynamic lubrication**, or **THL**. To begin with, the basic equations for thermohydrodynamic lubrication are described.

The usual Reynolds' equation is derived on the assumption that the coefficient of viscosity and density of the fluid are constant. In the case of thermohydrodynamic lubrication, however, both the coefficient of viscosity and the density change with temperature. Therefore, Reynolds' equation must be generalized so that these changes can be taken into account. This is the most important of the basic equations for thermohydrodynamic lubrication and is called the generalized Reynolds' equation.

In addition, the equation formulating the balance of the heat generated by shear in the fluid film, the heat carried away by convection and conduction, the heat accumulated in the fluid and so on is also an important basic equation. This is called the energy equation. Expressions for the temperature-dependence of the coefficient of viscosity and the density of the fluid are also necessary.

Besides the above equations, the equations of heat conduction within the solid parts such as the shaft and bearings, and that of heat transfer at the surface of solid parts are also required for the thermal analyses of a bearing. The thermal distortion of solid parts must sometimes be taken into consideration.

These equations are listed here. Among these, the generalized Reynolds' equation and the energy equation are explained in detail in the following sections.

1. Generalized Reynolds' equation for a lubricant film (see Section 8.2)
2. Energy equation for a lubricant film (see Section 8.3)
3. Equation of viscosity change of lubricant oil: $\mu = \mu_{in}\exp\{\beta(T_{in} - T)\}$
4. Equation of density change of lubricant oil: $\rho = \rho_{in}\{1 + \alpha_v(T_{in} - T)\}$
5. Equation of heat conduction inside solid parts: $\dfrac{\partial^2 T}{\partial x^2} + \dfrac{\partial^2 T}{\partial y^2} + \dfrac{\partial^2 T}{\partial z^2} = 0$
6. Equation of heat transfer at the surface of solid parts: $Q = h_c(T_s - T_a)$
7. Equation of heat expansion of solid parts: $\epsilon = \alpha(T - T_0)$

8.2 Generalized Reynolds' Equation

In the usual Reynolds' equation, it is assumed that the coefficient of viscosity and the density of the fluid are constant. This is equivalent to an assumption that the oil film temperature is uniform throughout the oil film. If a large amount of heat is generated in the oil film of a bearing, however, the change of oil film temperature in the sliding direction, in the film thickness direction, and in the direction normal to these cannot be ignored, and hence the change in coefficient of viscosity and density in these directions cannot be ignored either. A generalized Reynolds' equation that takes these changes into consideration was derived by D. Dowson as follows [3].

8.2.1 Balance of Forces

A small cubic element that is stationary in space is considered in a lubricant film and the balance of forces acting on the cube in the x and z directions is considered. Then the following equation is obtained in the same way as for Eq. 2.6 in Chapter 2:

$$\frac{\partial p}{\partial x} = \frac{\partial}{\partial y}\left(\mu \frac{\partial u}{\partial y}\right) \tag{8.1}$$

$$\frac{\partial p}{\partial z} = \frac{\partial}{\partial y}\left(\mu \frac{\partial w}{\partial y}\right) \tag{8.2}$$

where a Newtonian fluid is assumed. Further, although the coefficient of viscosity μ is a function of temperature, it can be considered to be a function of location in the case of stationary state:

$$\mu(x, y, z) \tag{8.3}$$

The pressure gradient $\partial p/\partial y$ in the film thickness direction is omitted here because it is usually very small.

8.2.2 Flow Velocity

First, the gradients of u and w in the film thickness direction are obtained by integrating Eq. 8.1 and Eq. 8.2 with respect to y:

$$\frac{\partial u}{\partial y} = \frac{\partial p}{\partial x}\frac{y}{\mu} + \frac{B(x,z)}{\mu} \tag{8.4}$$

$$\frac{\partial w}{\partial y} = \frac{\partial p}{\partial z}\frac{y}{\mu} + \frac{C(x,z)}{\mu} \tag{8.5}$$

where $B(x,z)$ and $C(x,z)$ are integral constants. Integrating the above equations once again with respect to y under the boundary conditions

$$u = U_1, \quad w = W_1 \quad \text{at} \quad y = 0$$
$$u = U_2, \quad w = W_2 \quad \text{at} \quad y = h$$

gives the following velocity components u and w:

$$u = U_1 + \frac{\partial p}{\partial x}\int_0^y \frac{y}{\mu}dy + B(x,z)\int_0^y \frac{dy}{\mu} \tag{8.6}$$

$$w = W_1 + \frac{\partial p}{\partial z}\int_0^y \frac{y}{\mu}dy + C(x,z)\int_0^y \frac{dy}{\mu} \tag{8.7}$$

where

$$B(x,z) = \frac{U_2 - U_1}{F_0} - \frac{F_1}{F_0}\frac{\partial p}{\partial x}$$

$$C(x,z) = \frac{W_2 - W_1}{F_0} - \frac{F_1}{F_0}\frac{\partial p}{\partial z}$$

$$F_0 = \int_0^h \frac{dy}{\mu}, \quad F_1 = \int_0^h \frac{ydy}{\mu}$$

8.2.3 Continuity Equation

The continuity equation for a compressible fluid is as follows:

$$\frac{\partial \rho}{\partial t} + \frac{\partial(\rho u)}{\partial x} + \frac{\partial(\rho v)}{\partial y} + \frac{\partial(\rho w)}{\partial z} = 0 \tag{8.8}$$

Integrating this in the direction of y from 0 to h and proceeding in the same way as for Eq. 2.14 in Chapter 2 yields the continuity equation for a lubricant film taking change of density into consideration as follows:

$$\int_0^h \frac{\partial \rho}{\partial t}dy + \frac{\partial}{\partial x}\int_0^h (\rho u)dy + \frac{\partial}{\partial z}\int (\rho w)dy$$
$$- (\rho U)_2 \frac{\partial h}{\partial x} - (\rho W)_2 \frac{\partial h}{\partial z} + \left[(\rho V)_2 - (\rho V)_1\right] = 0 \tag{8.9}$$

8.2.4 Generalized Reynolds' Equation

Let $\partial/\partial t = 0$, assuming a stationary state, and use the boundary conditions

$$V_1 = 0 \quad \text{at } y = 0, \quad U_2 = V_2 = W_2 = 0 \quad \text{at } y = h, \tag{8.10}$$

then, the continuity equation (Eq. 8.9) can be simplified as follows.

$$\frac{\partial}{\partial x}\int_0^h (\rho u)dy + \frac{\partial}{\partial z}\int (\rho w)dy = 0 \tag{8.11}$$

Subsituting u and w from Eqs. 8.6 and 8.7 into Eq. 8.11 leads to the generalized Reynolds' equation as follows:

$$\frac{\partial}{\partial x}\left[F_2\frac{\partial p}{\partial x}\right] + \frac{\partial}{\partial z}\left[F_2\frac{\partial p}{\partial z}\right] = \frac{\partial}{\partial x}\left[\left(F_4 - \frac{F_3}{F_0}\right)u\right]$$
$$+ \frac{\partial}{\partial z}\left[\left(F_4 - \frac{F_3}{F_0}\right)w\right] \tag{8.12}$$

where

$$F_0 = \int_0^h \frac{dy}{\mu}, \quad F_1 = \int_0^h \frac{y\,dy}{\mu} \tag{8.13}$$

$$F_2 = \frac{F_1}{F_0}F_3 - \int_0^h \rho \int_0^y \frac{y\,dy}{\mu}dy \tag{8.14}$$

$$F_3 = \int_0^h \rho \int_0^y \frac{dy}{\mu}dy, \quad F_4 = \int_0^h \rho\,dy \tag{8.15}$$

where the variation of μ not only in the x and z directions but also in the y direction is considered. Another form of generalized Reynolds' equation that is convenient when the variation of ρ in the y direction is not negligible is also proposed [8].

Equation 8.12 can be transformed into cylindrical coordinates (r, θ, z) as follows:

$$\frac{\partial}{\partial r}\left(F_2\frac{\partial p}{\partial r}\right) + \frac{1}{r^2}\frac{\partial}{\partial \theta}\left(F_2\frac{\partial p}{\partial \theta}\right) + \frac{F_2}{r}\frac{\partial p}{\partial r} = \omega\frac{\partial}{\partial \theta}\left(F_4 - \frac{F_3}{F_0}\right) \tag{8.16}$$

The r and θ components of the flow velocity in this case are as follows:

$$v_r = \left(\int_0^z \frac{z}{\mu}dz - \frac{F_1}{F_0}\int_0^z \frac{dz}{\mu}\right)\frac{\partial p}{\partial r} \tag{8.17}$$

$$v_\theta = r\omega\left(1 - \frac{1}{F_0}\int_0^z \frac{dz}{\mu}\right) + \left(\int_0^z \frac{z}{\mu}dz - \frac{F_1}{F_0}\int_0^z \frac{dz}{\mu}\right)\frac{1}{r}\frac{\partial p}{\partial \theta} \tag{8.18}$$

where F_i ($i = 0, 1, 2, 3, 4$) in the case of cylindrical coordinates are as follows, replacing y by z:

$$F_0 = \int_0^h \frac{dz}{\mu}, \qquad F_1 = \int_0^h \frac{z\,dz}{\mu} \qquad (8.19)$$

$$F_2 = \frac{F_1}{F_0} F_3 - \int_0^h \rho \int_0^z \frac{z\,dz}{\mu} dz \qquad (8.20)$$

$$F_3 = \int_0^h \rho \int_0^z \frac{dz}{\mu} dz, \qquad F_4 = \int_0^h \rho\,dz \qquad (8.21)$$

8.3 Energy Equation

To consider the temperature rise in a fluid under shear, the balance of the heat produced by viscous dissipation, the heat flow by convection and conduction, and the heat accumulated in the fluid must be investigated. An equation that describes such a balance of energy is called the energy equation. This is derived here according to Bird et al. [2].

8.3.1 General Energy Equation

A volume element stationary in space is considered in a flow of fluid as shown in Fig. 8.1. The following law of conservation of energy holds for the element.

[rate of increase of energy in the element (1)]
= [net inflow of energy by convection (2)]
+ [net inflow of energy by heat conduction (3)]
+ [work done by body force (gravity, etc.) on fluid (4)]
+ [work done by surface force (pressure) on fluid (5)]
+ [work done by surface force (stress) on fluid (6)]   (8.22)

It should be noted that the energy here includes internal energy and kinetic energy. If other forms of energy, such as chemical energy or electromagnetic energy, are involved, they will also be taken into consideration. Note that the internal energy (ρU per unit volume) is the energy of random microscopic motion of fluid molecules and is a function of temperature. The kinetic energy ($\rho V^2/2$ per unit volume) is the macroscopic kinetic energy of the fluid as a continuum. In this section, U is used for the internal energy and V for the general velocity. ρ is the density of the fluid.

If a two dimensional case is considered for simplicity, each term of the above equation (Eq. 8.22) can be written as follows:

$$(1) = +\Delta x \Delta y \frac{\partial}{\partial t}\left(\rho U + \frac{1}{2}\rho V^2\right)$$

$$(2) = -\Delta x \Delta y \left\{ \frac{\partial}{\partial x} u\left(\rho U + \frac{1}{2}\rho V^2\right) + \frac{\partial}{\partial y} v\left(\rho U + \frac{1}{2}\rho V^2\right) \right\}$$

$$(3) = -\Delta x \Delta y \left(\frac{\partial q_x}{\partial x} + \frac{\partial q_y}{\partial y} \right)$$

8.3 Energy Equation

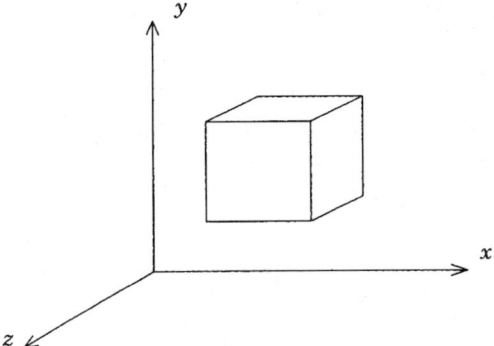

Fig. 8.1. Volume element in the fluid

$$(4) = + \Delta x \Delta y \, \rho \left(u g_x + v g_y \right)$$

$$(5) = - \Delta x \Delta y \left(\frac{\partial}{\partial x}(pu) + \frac{\partial}{\partial y}(pv) \right)$$

$$(6) = - \Delta x \Delta y \left\{ \frac{\partial}{\partial x}\left(\sigma_x u + \tau_{xy} v \right) + \frac{\partial}{\partial y}\left(\tau_{yx} u + \sigma_y v \right) \right\}$$

Substituting these into Eq. 8.22 above and dividing both sides by $\Delta x \Delta y$ yields the following equation:

$$\begin{aligned}
\frac{\partial}{\partial t}&\left(\rho U + \frac{1}{2}\rho V^2 \right) \\
= &- \left\{ \frac{\partial}{\partial x} u \left(\rho U + \frac{1}{2}\rho V^2 \right) + \frac{\partial}{\partial y} v \left(\rho U + \frac{1}{2}\rho V^2 \right) \right\} \\
&- \left(\frac{\partial q_x}{\partial x} + \frac{\partial q_y}{\partial y} \right) + \rho \left(u g_x + v g_y \right) - \left(\frac{\partial}{\partial x}(pu) + \frac{\partial}{\partial y}(pv) \right) \\
&- \left\{ \frac{\partial}{\partial x}\left(\sigma_x u + \tau_{xy} v \right) + \frac{\partial}{\partial y}\left(\tau_{yx} u + \sigma_y v \right) \right\}
\end{aligned} \quad (8.23)$$

This is a form of the energy equation. If the first term of the right-hand side is moved to the left side, the left side can be rewritten as follows:

$$\frac{\partial}{\partial t}\left(\rho U + \frac{1}{2}\rho V^2 \right) + \left\{ \frac{\partial}{\partial x} u \left(\rho U + \frac{1}{2}\rho V^2 \right) + \frac{\partial}{\partial y} v \left(\rho U + \frac{1}{2}\rho V^2 \right) \right\}$$

$$= \rho \frac{D}{Dt}\left(U + \frac{1}{2}V^2 \right) \quad (8.24)$$

where D/Dt is called the substantial derivative, and its meaning is as follows:

$$\frac{D}{Dt} = \frac{\partial}{\partial t} + u \frac{\partial}{\partial x} + v \frac{\partial}{\partial y} \quad (8.25)$$

168   8 Heat Generation and Temperature Rise

In the process of rearrangement, the following continuity conditions is used:

$$\frac{\partial \rho}{\partial t} + \frac{\partial (\rho u)}{\partial x} + \frac{\partial (\rho v)}{\partial y} = 0 \tag{8.26}$$

Equation 8.23 can be written as follows by using Eq. 8.24.

$$\rho \frac{D}{Dt}\left(U + \frac{1}{2}V^2\right)$$
$$= -\left(\frac{\partial q_x}{\partial x} + \frac{\partial q_y}{\partial y}\right) + \rho\left(ug_x + vg_y\right) - \left(\frac{\partial}{\partial x}(pu) + \frac{\partial}{\partial y}(pv)\right)$$
$$- \left\{\frac{\partial}{\partial x}\left(\sigma_x u + \tau_{xy} v\right) + \frac{\partial}{\partial y}\left(\tau_{yx} u + \sigma_y v\right)\right\} \tag{8.27}$$

Equations 8.23 and 8.27 are different forms of the energy equation, the former using a coordinate system at rest in the space, and the latter using a coordinate system moving with the fluid.

8.3.2 Energy Equation

The energy equation derived in the previous section is an equation for the sum of the internal energy and kinetic energy. If the terms relating to kinetic energy are subtracted from both sides of the equation, the equation for internal energy (thermal energy) will be obtained. This is often simply called the energy equation. Since this is very useful in practice, it will be derived below for the case of Eq. 8.27.

First, the kinetic energy of the fluid will be derived below. The equation of motion (Navier–Stokes equation) of the fluid in two dimensions can be written as follows:

$$\rho \frac{Du}{Dt} = -\frac{\partial p}{\partial x} - \left(\frac{\partial \sigma_x}{\partial x} + \frac{\partial \tau_{yx}}{\partial y}\right) + \rho g_x \tag{8.28}$$

$$\rho \frac{Dv}{Dt} = -\frac{\partial p}{\partial y} - \left(\frac{\partial \tau_{xy}}{\partial x} + \frac{\partial \sigma_y}{\partial y}\right) + \rho g_y \tag{8.29}$$

Multiplying both sides of the above two equations by u and v, respectively, yields the following equations:

$$\rho \frac{D}{Dt}\left(\frac{1}{2}u^2\right) = -u\frac{\partial p}{\partial x} - u\left(\frac{\partial \sigma_x}{\partial x} + \frac{\partial \tau_{yx}}{\partial y}\right) + \rho u g_x \tag{8.30}$$

$$\rho \frac{D}{Dt}\left(\frac{1}{2}v^2\right) = -v\frac{\partial p}{\partial y} - v\left(\frac{\partial \tau_{xy}}{\partial x} + \frac{\partial \sigma_y}{\partial y}\right) + \rho v g_y \tag{8.31}$$

Adding these two equations gives an equation of kinetic energy as follows:

$$\rho \frac{D}{Dt}\left(\frac{1}{2}V^2\right) = -\left(u\frac{\partial p}{\partial x} + v\frac{\partial p}{\partial y}\right)$$
$$- \left\{u\left(\frac{\partial \sigma_x}{\partial x} + \frac{\partial \tau_{yx}}{\partial y}\right) + v\left(\frac{\partial \tau_{xy}}{\partial x} + \frac{\partial \sigma_y}{\partial y}\right)\right\}$$
$$+ \rho\left(ug_x + vg_y\right) \tag{8.32}$$

8.3 Energy Equation

This can be rewritten as:

$$\rho \frac{D}{Dt}\left(\frac{1}{2}V^2\right) = -\left\{\frac{\partial}{\partial x}(pu) + \frac{\partial}{\partial y}(pv) - p\left(\frac{\partial u}{\partial x} + \frac{\partial v}{\partial y}\right)\right\}$$
$$-\left\{\frac{\partial}{\partial x}\left(\sigma_x u + \tau_{xy} v\right) + \frac{\partial}{\partial y}\left(\tau_{yx} u + \sigma_y v\right)\right.$$
$$\left. - \left(\sigma_x \frac{\partial u}{\partial x} + \tau_{xy}\frac{\partial v}{\partial x} + \tau_{yx}\frac{\partial u}{\partial y} + \sigma_y \frac{\partial v}{\partial y}\right)\right\}$$
$$+ \rho\left(ug_x + vg_y\right) \qquad (8.33)$$

Subtracting Eq. 8.33 from Eq. 8.27 gives the following equation for the internal energy:

$$\rho \frac{DU}{Dt} = -\left(\frac{\partial q_x}{\partial x} + \frac{\partial q_y}{\partial y}\right) - p\left(\frac{\partial u}{\partial x} + \frac{\partial v}{\partial y}\right)$$
$$-\left(\sigma_x \frac{\partial u}{\partial x} + \tau_{xy}\frac{\partial v}{\partial x} + \tau_{yx}\frac{\partial u}{\partial y} + \sigma_y \frac{\partial v}{\partial y}\right) \qquad (8.34)$$

Now, assuming Fourier's law, write heat fluxes q_x, q_y as follows:

$$q_x = -k_o \frac{\partial T}{\partial x} \qquad q_y = -k_o \frac{\partial T}{\partial y} \qquad (8.35)$$

where T is the temperature and k_o is the thermal conductivity. Further, assuming a Newtonian fluid, write the stress (excluding pressure) as follows [1]:

$$\sigma_x = \lambda\left(\frac{\partial u}{\partial x} + \frac{\partial v}{\partial y}\right) + 2\mu \frac{\partial u}{\partial x} \qquad (8.36)$$

$$\sigma_y = \lambda\left(\frac{\partial u}{\partial x} + \frac{\partial v}{\partial y}\right) + 2\mu \frac{\partial v}{\partial y} \qquad (8.37)$$

$$\tau_{xy} = \mu\left(\frac{\partial u}{\partial y} + \frac{\partial v}{\partial x}\right) \qquad (8.38)$$

$$\lambda + \frac{2}{3}\mu = 0 \qquad (8.39)$$

where μ is the coefficient of viscosity, λ is the secondary coefficient of viscosity, and $\lambda + \frac{2}{3}\mu$ is the volumetric coefficient of viscosity.

Substituting Eqs. 8.35 to 8.39 into Eq. 8.34 yields an energy equation as follows:

$$\rho \frac{DU}{Dt} = \frac{\partial}{\partial x}\left(k_o \frac{\partial T}{\partial x}\right) + \frac{\partial}{\partial y}\left(k_o \frac{\partial T}{\partial y}\right) - p\left(\frac{\partial u}{\partial x} + \frac{\partial v}{\partial y}\right) + \Phi \qquad (8.40)$$

where

$$\Phi = 2\mu\left[\left(\frac{\partial u}{\partial x}\right)^2 + \left(\frac{\partial v}{\partial y}\right)^2 + \frac{1}{2}\left(\frac{\partial u}{\partial y} + \frac{\partial v}{\partial x}\right)^2\right] + \lambda\left(\frac{\partial u}{\partial x} + \frac{\partial v}{\partial y}\right)^2 \qquad (8.41)$$

In the above equation, Φ is the energy that is converted to heat by viscous dissipation, and is called the dissipation energy.

Equations 8.40 and 8.41 were derived for the two-dimensional case. The energy equation and the dissipation energy for the three-dimensional case can be similarly obtained as follows:

$$\rho\frac{DU}{Dt} = \frac{\partial}{\partial x}\left(k_o\frac{\partial T}{\partial x}\right) + \frac{\partial}{\partial y}\left(k_o\frac{\partial T}{\partial y}\right) + \frac{\partial}{\partial z}\left(k_o\frac{\partial T}{\partial z}\right) - p\left(\frac{\partial u}{\partial x} + \frac{\partial v}{\partial y} + \frac{\partial w}{\partial z}\right) + \Phi \quad (8.42)$$

where

$$\Phi = 2\mu\left[\left(\frac{\partial u}{\partial x}\right)^2 + \left(\frac{\partial v}{\partial y}\right)^2 + \left(\frac{\partial w}{\partial z}\right)^2 + \frac{1}{2}\left(\frac{\partial u}{\partial y} + \frac{\partial v}{\partial x}\right)^2 + \frac{1}{2}\left(\frac{\partial v}{\partial z} + \frac{\partial w}{\partial y}\right)^2 + \frac{1}{2}\left(\frac{\partial w}{\partial x} + \frac{\partial u}{\partial z}\right)^2\right] + \lambda\left(\frac{\partial u}{\partial x} + \frac{\partial v}{\partial y} + \frac{\partial w}{\partial z}\right)^2 \quad (8.43)$$

In the energy equation (Eq. 8.42), the left-hand side is the convection term, the first three terms of the right-hand side are heat conduction terms, the fourth term is the compressibility term, and the last term is the viscous dissipation term.

8.3.3 Transformation of the Energy Equation

The energy equation is often more convenient if it is expressed in terms of temperature and specific heat rather than internal energy.

In the case of an incompressible fluid, the internal energy can be written as $U = c_v T$. Also, since there is no volume change in this case, the term $p(\cdots)$ in Eq. 8.42 and the term $\lambda(\cdots)^2$ in Eq. 8.43 are zero. Then, the energy equation (Eq. 8.42) will be as follows, provided that c_v is constant:

$$\rho c_v \frac{DT}{Dt} = \frac{\partial}{\partial x}\left(k_o\frac{\partial T}{\partial x}\right) + \frac{\partial}{\partial y}\left(k_o\frac{\partial T}{\partial y}\right) + \frac{\partial}{\partial z}\left(k_o\frac{\partial T}{\partial z}\right) + \Phi_v \quad (8.44)$$

where

$$\Phi_v = 2\mu\left[\left(\frac{\partial u}{\partial x}\right)^2 + \left(\frac{\partial v}{\partial y}\right)^2 + \left(\frac{\partial w}{\partial z}\right)^2 + \frac{1}{2}\left(\frac{\partial u}{\partial y} + \frac{\partial v}{\partial x}\right)^2 + \frac{1}{2}\left(\frac{\partial v}{\partial z} + \frac{\partial w}{\partial y}\right)^2 + \frac{1}{2}\left(\frac{\partial w}{\partial x} + \frac{\partial u}{\partial z}\right)^2\right] \quad (8.45)$$

In the case of a compressible fluid, we modify Eq. 8.42 by using the enthalpy $h = U + pv$ (v is volume, here) instead of the internal energy U. First, the continuity equation

$$\frac{\partial \rho}{\partial t} + \frac{\partial}{\partial x}(\rho u) + \frac{\partial}{\partial y}(\rho v) + \frac{\partial}{\partial z}(\rho w) = 0$$

can be modified as follows, by using the substantial derivative D/Dt:

$$\frac{D\rho}{Dt} + \rho\left(\frac{\partial u}{\partial x} + \frac{\partial v}{\partial y} + \frac{\partial w}{\partial z}\right) = 0$$

By using this, we rewrite $\rho(DU)/(Dt)$ as:

$$\rho\frac{DU}{Dt} = \rho\frac{D}{Dt}\left(h - \frac{p}{\rho}\right) = \rho\frac{Dh}{Dt} - \frac{Dp}{Dt} + \frac{p}{\rho}\frac{D\rho}{Dt}$$

$$= \rho\frac{Dh}{Dt} - \frac{Dp}{Dt} - p\left(\frac{\partial u}{\partial x} + \frac{\partial v}{\partial y} + \frac{\partial w}{\partial z}\right)$$

Substituting this into Eq. 8.42 yields an energy equation in terms of enthalpy as follows:

$$\rho\frac{Dh}{Dt} = \frac{\partial}{\partial x}\left(k_o\frac{\partial T}{\partial x}\right) + \frac{\partial}{\partial y}\left(k_o\frac{\partial T}{\partial y}\right) + \frac{\partial}{\partial z}\left(k_o\frac{\partial T}{\partial z}\right) + \frac{Dp}{Dt} + \Phi \quad (8.46)$$

The following relation is obtained from general relations for enthalpy:

$$\rho\frac{Dh}{Dt} = \frac{Dp}{Dt} - \alpha_v T\frac{Dp}{Dt} + \rho c_p\frac{DT}{Dt} \quad (8.47)$$

where α_v is the coefficient of cubic expansion. Therefore, Eq. 8.46 can be written as follows using specific heat and temperature instead of enthalpy:

$$\rho c_p\frac{DT}{Dt} = \frac{\partial}{\partial x}\left(k_o\frac{\partial T}{\partial x}\right) + \frac{\partial}{\partial y}\left(k_o\frac{\partial T}{\partial y}\right) + \frac{\partial}{\partial z}\left(k_o\frac{\partial T}{\partial z}\right) + \alpha_v T\frac{Dp}{Dt} + \Phi \quad (8.48)$$

In the case of a thin liquid lubrication film, the energy equation can be very much simplified. On the assumption that: (1) specific heat and thermal conductivity are constant, (2) gradients of flow velocities u and w in directions other than the film thickness direction can be neglected, and (3) thermal conduction in directions other than the film thickness direction can be ignored, Equations 8.44 and 8.48 can be written as follows:

$$\rho c_v\frac{DT}{Dt} = k_o\frac{\partial^2 T}{\partial y^2} + \mu\left\{\left(\frac{\partial u}{\partial y}\right)^2 + \left(\frac{\partial w}{\partial y}\right)^2\right\} \quad (8.49)$$

$$\rho c_p\frac{DT}{Dt} = k_o\frac{\partial^2 T}{\partial y^2} + \alpha_v T\frac{Dp}{Dt} + \mu\left\{\left(\frac{\partial u}{\partial y}\right)^2 + \left(\frac{\partial w}{\partial y}\right)^2\right\} \quad (8.50)$$

8.4 Temperature Distribution in Bearings

There are two ways for the heat generated in the lubricating film to leave the bearing. One way is convection. The heat is removed with the flow of fluid. The other is conduction. The heat flows inside the solid parts, such as the bearing and the shaft, and finally dissipates in the air. The total amount of heat that flows out by convection and conduction is equal to the total amount of heat generated.

In the case of a bearing, generally speaking, convection and conduction take place simultaneously and, as a result, the temperature distribution in a bearing is quite complicated. When a lubricating film is comparatively thick such as that of a bearing, heat is mainly carried away by convection, and when a lubricating film is very thin, as in the case of gears, heat mainly flows away by conduction.

Since it is quite difficult, theoretically or experimentally, to find the temperature distribution in a bearing, the problem was conventionally solved only after substantial simplifications. For example, the classical Reynolds' equation assumes constant viscosity, or equivalently assumes uniform temperature distribution in the oil film (temperature is the same everywhere) in the analysis. This is called **isoviscous anlysis**. In actual bearings, however, oil temperature is not uniform. Sometimes constant temperature or constant viscosity is assumed at least in the film thickness direction, conceding that they are not uniform in the other two directions. Such an analysis is called a **two-dimensional analysis**. In this case, it is not necessary to consider the heat conduction in pads, because there is no heat flow in the direction of the film thickness. The generalized Reynolds' equation and the energy equation will also be simplified. Strictly speaking, temperature distribution in an oil film is of course three dimensional, and so a **three-dimensional analysis** is necessary. A big problem in this case is that three-dimensional analyses take much computing time.

If the calculated results of isoviscous, two-dimensional, and three-dimensional analyses are not very much different, the easy calculation of isoviscous or two-dimensional analyses will be sufficient. In fact, however, there are marked differences between these results, as discussed below. It is therefore very important to carry out a three-dimensional analysis and to know the reality of the three-dimensional temperature distribution in bearings. Accurate prediction of various bearing characteristics, e.g., temperature distribution or the highest temperature, is very important in the design of a bearing, especially in the design of a new bearing. Simplified calculations may involve large errors in prediction of, for example, the highest temperature.

8.5 Temperature Analyses of Tilting Pad Thrust Bearings — Sector Pads

In this section, a sector pad of a tilting pad thrust bearing as shown in Fig. 8.2 is considered. A sector pad is supported by a pivot so that it can tilt in both the pitching direction and the rolling direction over a disk rotating at a constant speed. Let both the lubricating surfaces of the disk and the pad be rigid planes. The three-dimensional temperature distribution in the lubricant film and the pad, and also various characteristics of the bearing under the three dimensional condition are obtained by numerical computations. Subsequently, results are obtained also from two-dimensional and isoviscous analyses and compared with those of the three-dimensional analysis (Kim et al. [16] [20] [21]). These will be compared also with experiments (Kim et al. [23], Tanaka et al. [19]).

Similar studies include a two-dimensional analysis of a sector pad, taking distortion of the sector pad and the carry-over of lubricating oil into consideration (Ettles

[11]), a two-dimensional analysis of a sector pad taking into account the effect of centrifugal force on the lubricating oil (Pinkus and Lund [15]), a three-dimensional analysis of a fixed inclined rectangular pad bearing (Ezzat and Rhode [9]), and a three-dimensional analysis of a tilting sector-pad bearing taking into account only the inclination in pitch mode (Tieu [10]).

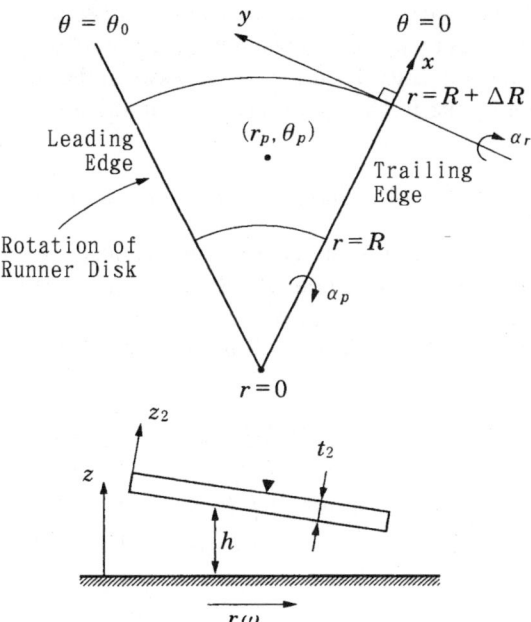

Fig. 8.2. Sector-type tilting pad bearing [20]

8.5.1 Basic Equations

A cylindrical coordinate system (r, θ, z) is considered with the origin at the center of rotation of the disk (see Fig. 8.2). Let the position of the pivot support be (r_p, θ_p). In the direction of pad thickness, a coordinate z_2 with the origin on the lubricating surface of the pad is used. Besides the usual assumptions of lubrication theory, the following assumptions are made:

1. The lubricating surfaces of the pad and the disk are rigid, and their thermal and elastic distortions are disregarded.
2. The velocity gradient and heat conduction in the r and θ directions can be ignored in comparison with those in the z direction.
3. The specific heat at constant pressure c_p and the coefficient of thermal expansion α_v of the lubricating oil are constant.

8 Heat Generation and Temperature Rise

4. The thermal conductivity of the lubricating oil k_o and the thermal conductivity of the pad k_s are constant.
5. The coefficient of viscosity μ and the density ρ of the lubricating oil are functions of temperature T only.

The basic equations in this case are stated below.

Let the state when the lubricating surfaces of the pad and the disk are parallel to each other and their separation is h_{0R} be the standard state. Assume that the pad rotates around the x axis by α_p (pitching angle) and around the y axis by α_r (rolling angle) from the standard state. Then the gap h between the pad and the disk (film thickness) is given by the following equation, assuming a small inclination of the pad, $\alpha_p \ll 1$ and $\alpha_r \ll 1$:

$$h(r, \theta) = (R + \Delta R - r\cos\theta)\alpha_r + \alpha_p r \sin\theta + h_{0R} \tag{8.51}$$

Inclinations of the pad α_p and α_r are actually automatically determined by the balance of the oil film force. The minimum film thickness h_0 appears at $(R+\Delta R, 0)$ when α_r is positive and at $(R, 0)$ when α_r is negative.

The flow velocity of the lubricating oil in the radial and the circumferential directions is:

$$v_r = \left(\int_0^z \frac{z}{\mu}dz - \frac{F_1}{F_0}\int_0^z \frac{dz}{\mu}\right)\frac{\partial p}{\partial r} \tag{8.52}$$

$$v_\theta = -r\omega\left(1 - \frac{1}{F_0}\int_0^z \frac{dz}{\mu}\right) + \left(\int_0^z \frac{z}{\mu}dz - \frac{F_1}{F_0}\int_0^z \frac{dz}{\mu}\right)\frac{1}{r}\frac{\partial p}{\partial \theta} \tag{8.53}$$

where

$$F_0 = \int_0^h \frac{dz}{\mu}, \quad F_1 = \int_0^h \frac{zdz}{\mu}$$

The generalized Reynolds' equation is written as:

$$\frac{\partial}{\partial r}\left(F_2 \frac{\partial p}{\partial r}\right) + \frac{1}{r^2}\frac{\partial}{\partial \theta}\left(F_2 \frac{\partial p}{\partial \theta}\right) + \frac{F_2}{r}\frac{\partial p}{\partial r} = \omega \frac{\partial}{\partial \theta}\left(F_4 - \frac{F_3}{F_0}\right) \tag{8.54}$$

where

$$F_2 = \frac{F_1}{F_0}F_3 - \int_0^h \rho \int_0^z \frac{zdz}{\mu}dz,$$

$$F_3 = \int_0^h \rho \int_0^z \frac{dz}{\mu}dz, \quad F_4 = \int_0^h \rho dz$$

The energy equation for the oil film is:

$$c_p\rho\left(v_r\frac{\partial T}{\partial r} + \frac{v_\theta}{r}\frac{\partial T}{\partial \theta} + v_z\frac{\partial T}{\partial z}\right)$$

$$= k_o\frac{\partial^2 T}{\partial z^2} + \alpha_v T\left(v_r\frac{\partial p}{\partial r} + \frac{v_\theta}{r}\frac{\partial p}{\partial \theta}\right) + \mu\left\{\left(\frac{\partial v_r}{\partial z}\right)^2 + \left(\frac{\partial v_\theta}{\partial z}\right)^2\right\} \tag{8.55}$$

8.5 Temperature Analyses of Tilting Pad Thrust Bearings — Sector Pads

The heat conduction equation for the pad is:

$$\frac{\partial^2 T_2}{\partial r^2} + \frac{1}{r^2}\frac{\partial^2 T_2}{\partial \theta^2} + \frac{\partial^2 T_2}{\partial z_2^2} + \frac{1}{r}\frac{\partial T_2}{\partial r} = 0 \qquad (8.56)$$

The coefficient of viscosity of the lubricating oil is:

$$\mu = \mu_{in} \exp\{\beta (T_{in} - T)\} \qquad (8.57)$$

The density of the lubricating oil is:

$$\rho = \rho_{in}\{1 + \alpha_v(T_{in} - T)\} \qquad (8.58)$$

8.5.2 Boundary Conditions

As boundary conditions for the generalized Reynolds' equation, it is assumed that the pressure around the pad is atmospheric pressure, i.e.,

$$p(r, 0) = p(r, \theta_0) = p(R, \theta) = p(R + \Delta R, \theta) = 0 \qquad (8.59)$$

As boundary conditions for the energy equation of the oil film, it is assumed first that both the inlet oil temperature and the disk surface temperature are equal to T_{in}, i.e.,

$$T(r, \theta_0, z) = T(r, \theta, 0) = T_{in} \qquad (8.60)$$

It is also assumed that the oil film temperature changes parabolically in the r direction on the inner and outer circular boundaries of the pad. Furthermore, from the continuity of the heat flux at the interface of the pad and the oil film in the direction normal to the interface, it is assumed that:

$$k_o \left(\frac{\partial T}{\partial z}\right)_{z=h} = k_s \left(\frac{\partial T_2}{\partial z_2}\right)_{z_2=0} \qquad (8.61)$$

where T and T_2 are the temperature of the lubricating oil and the pad, respectively, and k_o and k_s are the thermal conductivities of the lubricating oil and the pad, respectively.

As boundary conditions for the heat conduction equation of the pad, it is assumed that Eq. 8.61 holds at the lubricating surface and that heat flux is continuous at the other surfaces, the ambient temperature T_a and the heat transfer coefficient at the interfaces h_c being given.

8.5.3 Numerical Analyses

Before numerical computations, each variable is nondimensionalized as follows. Since both the lubricating film and the pad become a cube of side 1, the forms of the computational domain in the finite difference method become very simple:

8 Heat Generation and Temperature Rise

$$\bar{r} = \frac{r-R}{\Delta R}, \quad \bar{\theta} = \frac{\theta}{\theta_0}, \quad \bar{z} = \frac{z}{h}, \quad \bar{h} = \frac{h}{h_0}$$

$$\bar{v}_r = \frac{v_r}{R\omega}, \quad \bar{v}_\theta = \frac{v_\theta}{r\omega}, \quad \bar{v}_z = \frac{v_z}{R\omega}, \quad \bar{z}_2 = \frac{z_2}{t_2}$$

$$\bar{\mu} = \frac{\mu}{\mu_{in}}, \quad \bar{\rho} = \frac{\rho}{\rho_{in}}, \quad \bar{T} = \frac{T}{T_{in}}, \quad \bar{T}_2 = \frac{T_2}{T_{in}}$$

$$\bar{p} = \frac{p h_0^2}{R^2 \omega \mu_{in}} \tag{8.62}$$

Then, the lubricating film and the pad are divided into, for example, 20 divisions in the r and θ directions, and into, for example, 15 divisions in the z direction, and Eqs. 8.51 to 8.58 are discretized for the finite difference analysis. The centered difference is used, except in the case of $\partial T/\partial \theta$ in the energy equation (Eq. 8.55) for which the backward difference is used. This is solved by the successive overre-

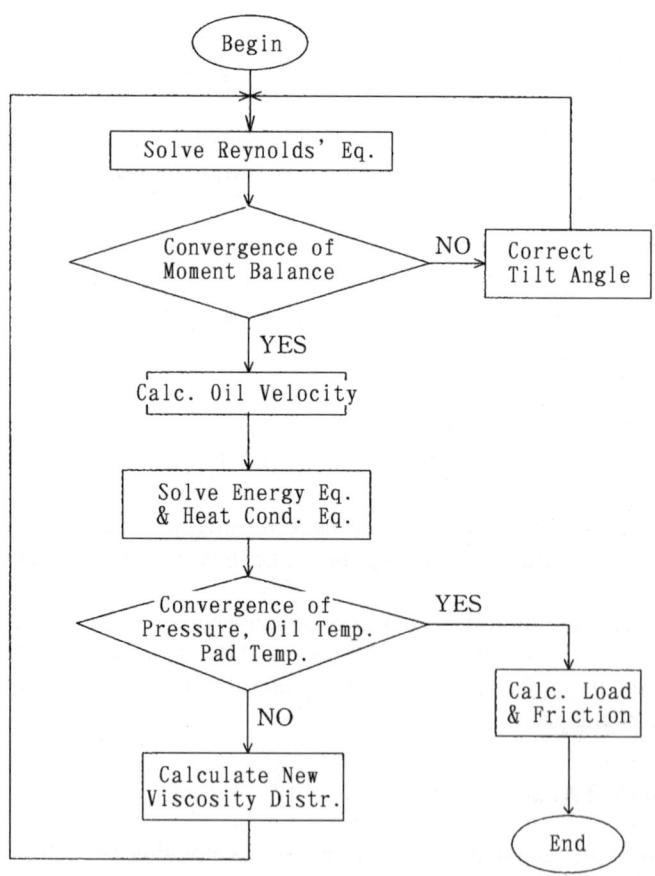

Fig. 8.3. Flow chart for computation [20]

8.5 Temperature Analyses of Tilting Pad Thrust Bearings — Sector Pads 177

laxation method (SOR method), following the flow chart of Fig. 8.3. Calculation is repeated until the error from the last calculated value becomes, for example, smaller than 10^{-5} at all nodal points. Further, to make the center of pressure coincide with the pivot position, the inclination of the pad is adjusted by the Newton–Raphson calculation.

The constants in Table 8.1 were used in the following calculations.

Table 8.1. Specifications of the sector pad bearing and the oil

Inner radius of pad	$R = 1.6 \times 10^{-1}$ m
Radial extent of pad	$\Delta R = 7.7 \times 10^{-2}$ m
Angular extent of pad	$\theta_0 = \pi/8$ rad
Thickness of pad	$t = 2.5 \times 10^{-2}$ m
Thermal conduction of pad	$k_s = 1.08 \times 10^2$ W/m°C
heat transfer at surface of pad	$h_c = 5.81 \times 10^2$ W/m² °C
Viscosity of oil (27°C)	$\mu = 5.0 \times 10^{-2}$ Pa · s
Viscosity index of oil	$\beta = 4.91 \times 10^{-2}$ °C^{-1}
Specific heat of oil	$c_p = 2.09 \times 10^3$ J/kg°C
Thermal conductivity of oil	$k_o = 0.214$ W/m °C
Coefficient of thermal expansion of oil	$\alpha_v = 7.34 \times 10^{-4}$ °C^{-1}
Density of oil (27°C)	$\rho = 8.55 \times 10^2$ kg/m³

8.5.4 Examples of Three-Dimensional Analyses of Temperature Distribution

Figure 8.4 shows isothermal lines for the lubricating surface of the pad and for the central cross section of the lubricant film and the pad, parameters being $h_0 = 100\,\mu$m, $N = 3000$ rpm, $T_{in} = T_a = 47°$C, the pivot position in nondimensional radial coordinates $\bar{r}_p = (r_p - R)/\Delta R = 0.51$ and in nondimensional circumferential coordinates $\bar{\theta}_p = \theta_p/\theta_0 = 0.39$.

On the lubricating surface of the pad, temperature rises in the circumferential direction gradually near the leading edge and then quicker near the trailing edge. The highest temperature is found on the trailing edge near the outer radius. The temperature does not change very much in the radial direction. The shapes of the isothermal lines are very close to the experimental results under the same conditions (the same pivot position, and so forth) [23].

The highest temperature in Fig. 8.4 is approximately 58°C, which is 11°C higher than the entrance oil temperature. Generally speaking, the highest temperature falls if the heat transfer coefficient h_c of the pad surface is increased; however, the fall in highest temperature is only 2° – 4°C or so even if h_c is increased by $10-10^2$ times [16]. This probably shows that most of the heat is carried away by the convection of the lubricating oil.

In the central cross section of the lubricant film and the pad, the temperature in the lubricant film (the wedge-shaped domain in the lower half; the thickness is

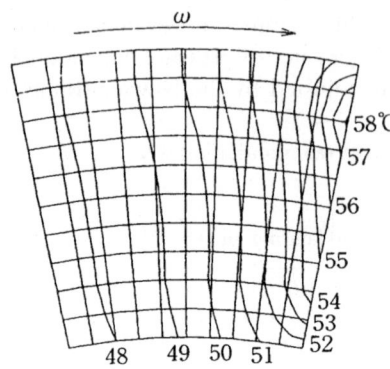 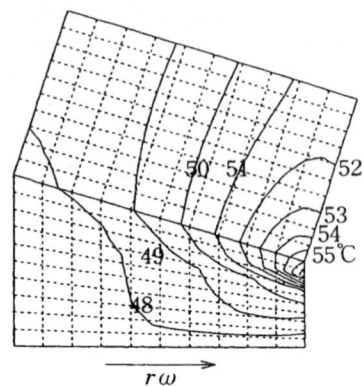

Fig. 8.4. Temperature distribution on the lubricating surface (*left*) and the central cross section of the lubricant film (*lower right*) and pad (*upper right*) [16]

exaggerated) rises gradually from the disk surface, and rises quickly near the pad surface. The highest temperature is found at the trailing edge of the lubricant film near the pad surface. The temperature distribution in the pad (the rectangular domain in the upper half) is as shown in the figure.

8.5.5 Comparisons of Three-Dimensional, Two-Dimensional, and Isoviscous Analyses

Figure 8.5a-c shows comparisons of the nondimensional load capacity, the temperature rise (= highest temperature - entrance temperature), and the nondimensional frictional torque calculated by three-dimensional, two-dimensionalal and isoviscous analyses, under the conditions $T_{in} = T_d = T_a = 27°C$, $h_0 = 100$ μm, and $N = 1500$ rpm. The horizontal axis shows the nondimensional circumferential coordinates of the pivot position $\bar{\theta}_p = \theta_p/\theta_0$, and the parameters in the figure are the nondimensional radial coordinates of the pivot position $\bar{r}_p = (r_p - R)/\Delta R$.

The nondimensional load capacity in Fig. 8.5a-ca shows that the three-dimensional analyses give the lowest load capacity, the two-dimensional analyses give an intermediate figure, and the isoviscous analyses give the highest load capacities. Fig. 8.5a also shows that there is an optimal pivot position $\bar{\theta}_p$ that gives the highest load capacity in each of the three analyses, $\bar{\theta}_p$ being slightly dependent on $\bar{r}_p$. The load capacity, which changes considerably with $\bar{r}_p$, becomes maximum when $\bar{r}_p = 0.51$ for any method of analysis here. The load capacity depends not only on the pad inclination but also on the oil viscosity, which is a function of temperature and hence of location. This affects the optimal pivot position in a complicated way. One must be aware that the isoviscous and the two-dimensional analyses may give load capacity estimates that are too high.

As for the temperature rise shown in Fig. 8.5a-cb, the two-dimensional analysis gives a value considerably lower than the three-dimensional analysis does, although

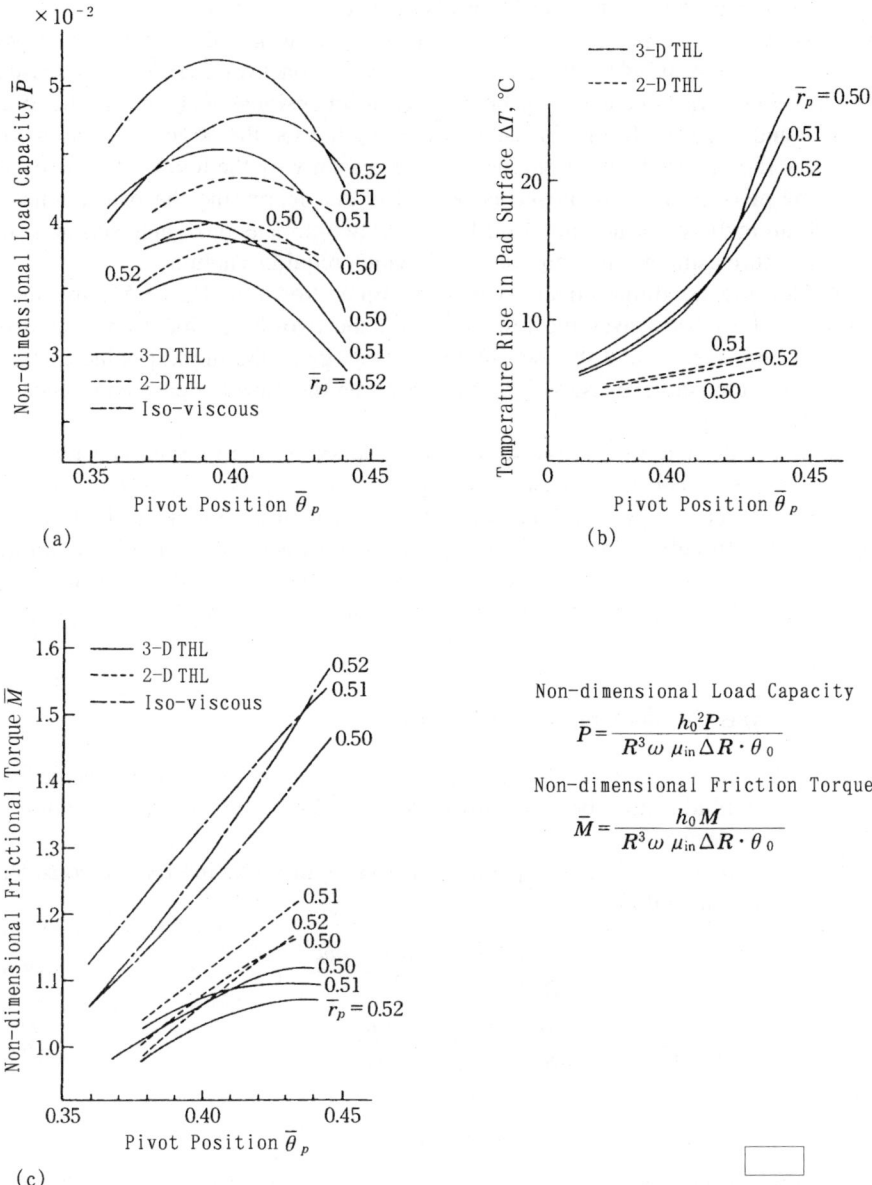

Fig. 8.5a-c. Comparisons of bearing characteristics using three-dimensional (*3-D*), two-dimensional (*2-D*), and isoviscous analyses [20]. **a** nondimensional load capacity, **b** temperature rise, **c** nondimensional frictional torgue. $\bar{r}_p$, nondimensional radial coordinates of the pivot position

180   8 Heat Generation and Temperature Rise

it does not consider the heat flow into solid surfaces. The reason for this is that only the average temperature is considered in the film thickness direction in two-dimensional analysis. If the pivot position is moved backward (if $\bar{\theta}_p$ is decreased), the temperature rise decreases for both the two-dimensional analysis and the three-dimensional analysis. In the three-dimensional analysis, the large temperature rise falls very rapidly as shown in the figure. The reason why the temperature rise falls when the pivot position is moved backward is that increasing the pad inclination leads to an increase in the rate of oil flow. The two-dimensional analysis can result in temperature estimates that are too small, which requires caution.

As for the nondimensional frictional torque shown in Fig. 8.5c, the three-dimensional analysis gives the lowest values, the two-dimensional analysis gives values a little higher, and the isoviscous analysis gives the highest values. This is because the temperature rise is larger in the three-dimensional case and hence the viscosity is lower.

It should be noted here that the results obtained by the three-dimensional, the two-dimensional, and the isoviscous analyses differ considerably in terms of load capacity, temperature rise, and frictional torque. Therefore, the analysis of a bearing should originally be performed using the three-dimensional analysis; and other approaches cannot substitute for it. When performing the two-dimensional or the isoviscous analyses, it should be kept in mind that they are considerable approximations.

8.5.6 Analysis Considering Inertia Forces

In high speed tilting pad thrust bearings, the influence of the inertia force (centrifugal force) on the fluid cannot be disregarded (Kim et al. [21]). Reynolds' equation for such a case is derived here.

In reference to Fig. 8.2, the equations of balance that take inertia into consideration are written as follows:

$$\frac{\partial}{\partial z}\left(\mu \frac{\partial v_r}{\partial z}\right) = \frac{\partial p}{\partial r} - \frac{\rho v_\theta^2}{r} \qquad (8.63)$$

$$\frac{\partial}{\partial z}\left(\mu \frac{\partial v_\theta}{\partial z}\right) = \frac{1}{r}\frac{\partial p}{\partial \theta} \qquad (8.64)$$

$$\frac{\partial p}{\partial z} = 0 \qquad (8.65)$$

The second term on the right-hand side of Eq. 8.63 is an inertia term. Integrating Eq. 8.64 under the boundary condition

$$v_\theta = r\omega \text{ at } z = 0, \quad v_\theta = 0 \text{ at } z = h$$

yields:

$$v_\theta = r\omega + \frac{1}{r}\frac{\partial p}{\partial \theta}\int_0^z \frac{z}{\mu}dz - \left(\frac{r\omega}{F_0} + \frac{F_1}{F_0}\frac{1}{r}\frac{\partial p}{\partial \theta}\right)\int_0^z \frac{dz}{\mu} \qquad (8.66)$$

8.5 Temperature Analyses of Tilting Pad Thrust Bearings — Sector Pads

Subsituting this into Eq. 8.63 and integrating under the boundary conditions

$$v_r = 0 \text{ at } z = 0 \text{ and } z = h$$

gives the following equation:

$$v_r = \frac{\partial p}{\partial r} \int_0^z \frac{z}{\mu} dz + \left(\frac{1}{r} \frac{G_0}{F_0} - \frac{F_1}{F_0} \frac{\partial p}{\partial r} \right) \int_0^z \frac{dz}{\mu} - \frac{1}{r} \int_0^z \left[\frac{1}{\mu} \int_0^z \rho(v_\theta)^2 dz \right] dz \quad (8.67)$$

where F_0, F_1, and F_2 are as follows:

$$F_0 = \int_0^h \frac{dz}{\mu}, \quad F_1 = \int_0^h \frac{z}{\mu} dz, \quad G_0 = \int_0^h \left[\frac{1}{\mu} \int_0^z \rho(v_\theta)^2 dz \right] dz \quad (8.68)$$

On the other hand, integration of the continuity equation with respect to z from 0 to h, the use of

$$v_z = 0 \text{ at } z = 0, \quad v_r = v_\theta = v_z = 0 \text{ at } z = h$$

and use of the formula for the change of order of differentiation and integration yields:

$$r \frac{\partial}{\partial r} \int_0^h \rho v_r dz + \int_0^h \rho v_r dz + \frac{\partial}{\partial \theta} \int_0^h \rho v_\theta dz = 0 \quad (8.69)$$

Substituting Eqs. 8.66 and 8.67 into Eq. 8.69 yields the following Reynolds' equation, which takes the inertia force and three-dimensional temperature change into consideration, as follows:

$$\frac{\partial}{\partial r} \left(G_1 \frac{\partial p}{\partial r} + G_2 \right) + \frac{1}{r} \left(G_1 \frac{\partial p}{\partial r} + G_2 \right) + \frac{1}{r^2} \frac{\partial}{\partial \theta} \left(G_1 \frac{\partial p}{\partial \theta} \right) = \frac{1}{r} \frac{\partial G_3}{\partial \theta} \quad (8.70)$$

where G_1, G_2, and G_3 are given as follows:

$$G_1 = \int_0^h \rho \left[\int_0^z \frac{z}{\mu} dz - \frac{F_1}{F_0} \int_0^z \frac{dz}{\mu} \right] dz$$

$$G_2 = \int_0^h \rho \left[\frac{1}{r} \frac{G_0}{F_0} \int_0^z \frac{dz}{\mu} - \frac{1}{r} \int_0^z \left\{ \frac{1}{\mu} \int_0^z \rho(v_\theta)^2 dz \right\} dz \right] dz$$

$$G_3 = \int_0^h \rho r \omega \left[\frac{1}{F_0} \int_0^z \frac{dz}{\mu} - 1 \right] dz \quad (8.71)$$

By means of Eq. 8.70, the three-dimensional thermohydrodynamic lubrication analysis that takes the inertia force into consideration can be carried out. The procedure of numerical computation is the same as before. When $\mu = \rho = $ constant, Eq. 8.70 coincides with the Reynolds' equation (Eq. 4.34) derived in Chapter 4 with reference to cylindrical coordinates.

182   8 Heat Generation and Temperature Rise

Examples of the three-dimensional analysis of thermohydrodynamic lubrication taking the inertia forces into consideration are shown below. Specifications of the bearing and the lubricating oil are listed in Table 8.1.

Figure 8.6 shows the calculated temperature rise, and illustrates how the difference between the highest and the lowest temperature (entrance temperature) on the pad surface, ΔT, changes with the rotational speed N. The operating conditions of the bearing are: $\bar{r}_p = 0.51$, $\bar{\theta}_p = 0.42$, $h_0 = 100\ \mu m$, and $T_{in} = 47°C$. If the inertia term is taken into consideration, ΔT is calculated to be larger than that for the case where no inertia effect is considered, and the larger N is, the larger ΔT is. This difference is due to the effect of the velocity component produced in the radial direction.

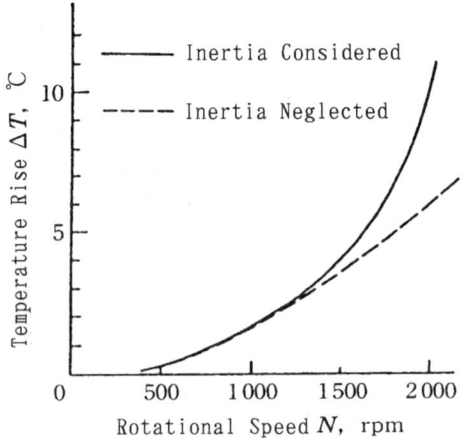

Fig. 8.6. Temperature rise [21]

Figure 8.7 shows the relation between the nondimensional load capacity $\bar{P}$ (see Fig. 8.5a-c) and N under the same operating conditions. If the inertia term is ignored, $\bar{P}$ decreases monotonously along with the increase in N. This is due to the decrease in viscosity along with the temperature rise in the oil film shown in Fig. 8.6. On the other hand, if an inertia term is taken into consideration, the calculated load capacity tends to be larger due to the flow velocity in the radial direction. As a result, in the area where the curve is upward convex in Fig. 8.7, the load capacity $\bar{P}$ is about 10% higher than that obtained ignoring the inertia term. If the rotating speed exceeds 2000 rpm, however, the load capacity falls because the influence of the decrease in viscosity due to temperature rise exceeds that of the inertia.

Figure 8.8 shows the relation between the inclination (tilt) of a pad and the rotating speed N under the same operating conditions. $\bar{\alpha}_p$ and $\bar{\alpha}_r$ are the circumferential inclination α_p and the radial direction inclination α_r of the pad multiplied by $\Delta R/h_0$, respectively. Even though both of the inclinations increase with the increase in N,

8.5 Temperature Analyses of Tilting Pad Thrust Bearings — Sector Pads

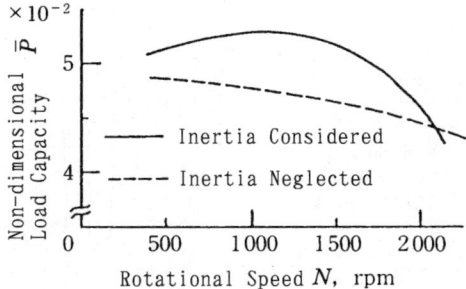

Fig. 8.7. Nondimensional load capacity [21]

irrespective of the existence of an inertia force, the inclinations increase more when an inertia force is taken into consideration, and in that case the values of $\bar{\alpha}_p$ and $\bar{\alpha}_r$ are also large. These differences are also due to the velocity component in the radial direction.

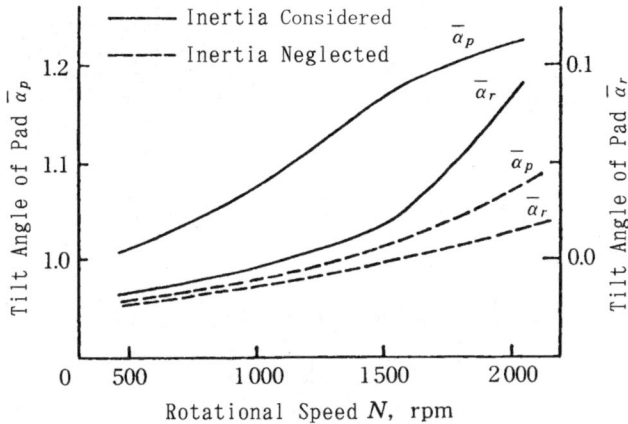

Fig. 8.8. Inclination of the pad [21]

Figure 8.9 shows examples of the calculated temperature distributions on the lubricating surface of the pad. Considering the inertia term increases both the temperature rise and the temperature gradient in the radial direction.

Figure 8.10 shows the pressure distribution in the oil film. Considering the inertia force increases the pressure generated to some extent. However, since the inclination always changes so that the position of the pressure center and that of the pivot position coincide, the positions of the pressure distribution curves and that of the highest pressure hardly change, whether the inertia force is considered or not.

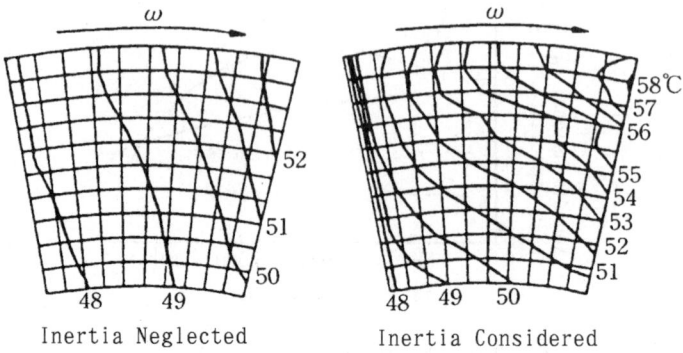

Fig. 8.9. Temperature distribution on the lubricating surface of a pad [21]

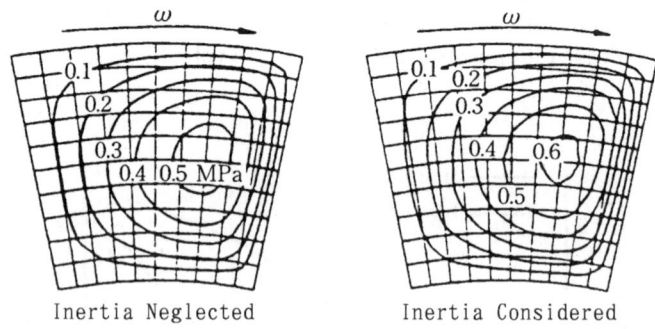

Fig. 8.10. Pressure distribution in an oil film [21]

8.5.7 Comparison of Calculated Results and Experiments

In an experiment on a sector pad of outside radius 237 mm, inner radius 160.7 mm, angular extent 22.5° and thickness 25 mm, temperatures were measured with 41 thermocouples embedded at a depth of 0.5 mm from the lubricating surface and were compared with theoretical results (Kim et al. [23]). Pivot positions were $\bar{r}_p = 0.516$ and $\bar{\theta}_p = 0.436$ in nondimensional coordinates.

Figure 8.11 shows the relationships between the average pad pressure p_m and the highest temperature on the pad surface T_{max} determined by two-dimensional and three-dimensional analyses, together with experimental results for comparison. Whereas the results of the two-dimensional analysis are quite different from the experimental results, those of the three-dimensional analysis are close to the experimental values, which shows the necessity of performing the three-dimensional analysis.

Figure 8.13a-c shows the relationship between the rotational speed of the disk N and the highest temperature on the pad surface T_{max} determined by analyses that

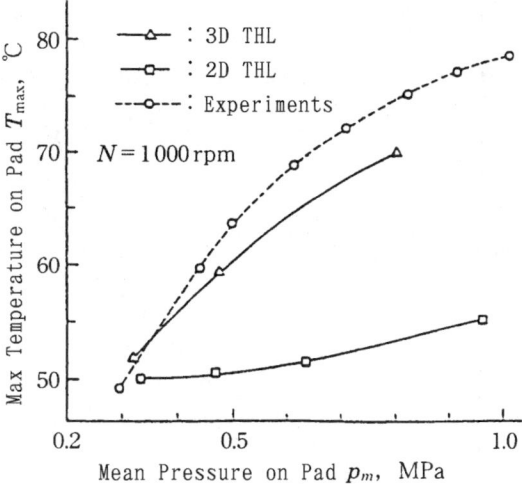

Fig. 8.11. Average bearing pressure and highest temperature on the pad surface [23]. *THL*, thermohydrodynamic lubrication

include or neglect the inertia term, together with experimental resutls for comparison. For high rotating speeds, it turns out that the analysis that includes the inertia term fits the experimental results better.

Figure 8.13a-c shows examples of the measured temperature distribution on the lubricating surface of a pad. The rotating speed in Figs. 8.13a-cb,c is twice that in Fig. 8.13a-ca, and the average pressure in Fig. 8.13a-cc is twice that in Figs. 8.13a-ca,b. In these figures, the maximum temperature is naturally high when the rotational speed is high, and the temperature gradient in the radial direction also tends to be large in this case. This tendency qualitatively resembles that shown in Fig. 8.9. Further, comparisons of Figs. 8.13a-cb,c show that high bearing pressures yield high temperature rises and that the high temperature domain tends to move toward the leading edge of the pad.

8.6 Temperature Analyses of Circular Journal Bearings

In this section, a perfectly circular journal bearing with an axial oil groove at the top of the bearing is considered. The oil groove extends almost the full length (width) of the bearing (see Fig. 8.14 and Table 8.2). Theoretical analyses of the temperature distribution within the oil film and the bearing metal are carried out taking the mixture of supply oil and circulation oil and the influence of oil film rupture into consideration. Detailed experiments using 144 thermocouples were also conducted to measure the temperature distribution within the bearing metal, and the results are compared with the above-mentioned theory (Mitsui et al. [17] [22]).

186   8 Heat Generation and Temperature Rise

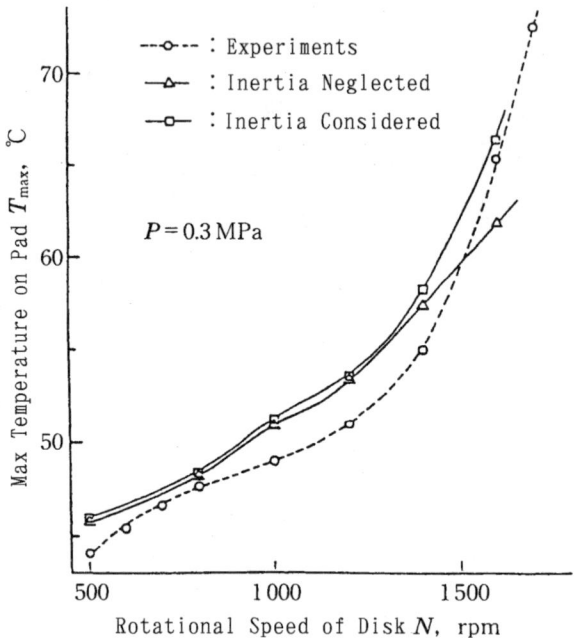

Fig. 8.13a-c. Rotating speed of the disk and the highest temperature on the pad surface [23]

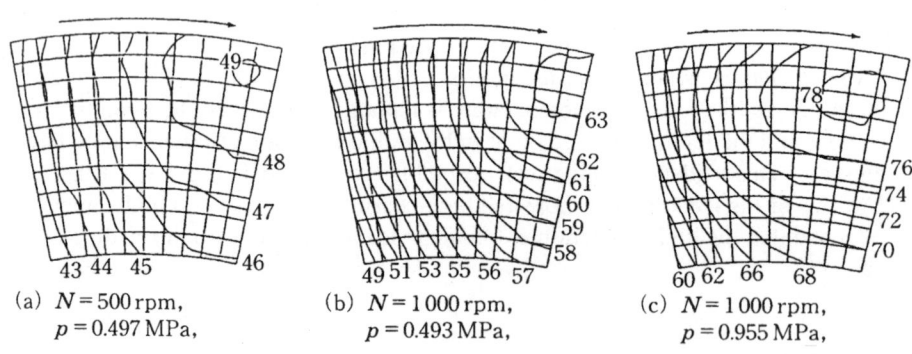

(a) $N = 500$ rpm,
$p = 0.497$ MPa,

(b) $N = 1000$ rpm,
$p = 0.493$ MPa,

(c) $N = 1000$ rpm,
$p = 0.955$ MPa,

Fig. 8.13a-c. Measured temperature distribution on the pad surface [23]. $\bar{r}_p = 0.516$, $\bar{\theta}_p = 0.436$, $T_{in} = 37°C$

8.6 Temperature Analyses of Circular Journal Bearings

Although not many experimental studies on the thermal characteristics of circular journal bearings have been reported, pioneering work by Dowson et al. [4] and also that by Woolacott and Cooke [6] can be mentioned as those that pursue the temperature characteristics of bearings by precise experiments. Theoretical investigations are more common, and those of particular note inculde research by Dowson and March, who solved Reynolds' equation and the energy equation in the case of an infinitely long bearing on the basis of the above experimental research [5]; research by McCallion et al., who solved Reynolds' equation, the energy equation, and the heat conduction equation of bearing metal, but the two former equations were treated separately [7]; and research by Boncompain, Frene, et al., who solved these three equations simultaneously [14] [18]. However, there has been no paper on perfect circular bearings that takes full account of the mixing of the supply oil and circulation oil, the influence of oil film rupture, and so on. In the case of partial bearings, papers by Suganami et al. can be mentioned [12] [13].

8.6.1 Basic Equations

As basic equations for a lubricant film, the equations for the fluid velocity, the generalized Reynolds' equation, and the energy equation are considered. These equations are conveniently written in rectangular coordinates, the lubricant film being developed on a plane. In solving these equations by the finite difference method, nondimensionalization of variables in a way similar to that used in the previous section is recommended.

Changes in the density of the lubricating oil will be disregarded and the following equation is assumed for the coefficient of viscosity:

$$\mu = \mu_{\text{in}} \exp\{\beta \, (T_{\text{in}} - T)\} \tag{8.72}$$

As a heat conduction equation within the bearing metal, the following equation in cylindrical coordinates will be used, T_b being the temperature (the temperature gradient in the axial direction is assumed to be negligible):

$$\frac{\partial^2 T_b}{\partial r^2} + \frac{1}{r} \frac{\partial T_b}{\partial r} + \frac{1}{r^2} \frac{\partial^2 T_b}{\partial \phi^2} = 0 \tag{8.73}$$

An infinite series solution, not a numerical solution, is used for this equation.

8.6.2 Boundary Conditions

In circular journal bearings, there are many complicated problems, such as rupture of the oil film in the negative pressure region, and mixing of the supply oil (newly supplied oil) and the recirculating oil (oil that is returned after circulation in the bearing) near the oil groove. The following boundary conditions are considered here.

As boundary conditions for pressure, it is assumed that the pressure is zero at both ends of the bearing and that the pressure obeys Reynolds' boundary condition at the end of the oil film in the circumferential direction.

188   8 Heat Generation and Temperature Rise

It is observed in the negative pressure region that the lubricant film contracts as shown in Fig. 8.14 and splits into several or many streams. Assuming that the oil film splits into a sufficiently large number of streams, the energy equation is applied to these streams in the negative pressure region.

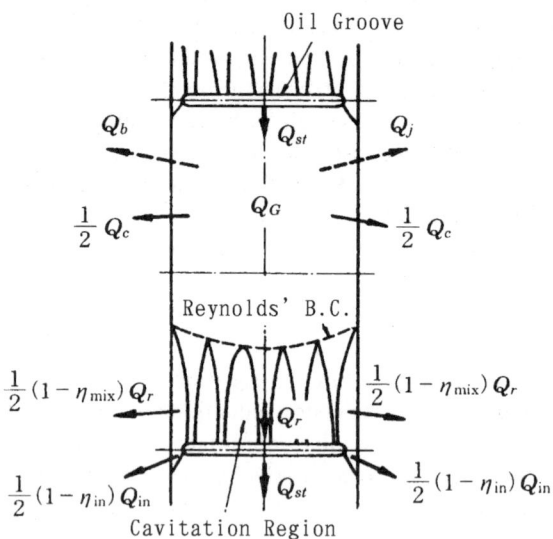

Fig. 8.14. Oil flow and heat flow in a circular journal bearing. *BC*, boundary condition [17]

As the boundary conditions for temperature, it is assumed that the temperature of the oil immediately downstream of the oil groove is equal to that of the mixture of the supply oil and the recirculated oil, that the temperature and the heat flux are continuous at the interface of the oil and the inner surface of the bearing metal, and that the heat flux of heat transfer is continuous at the interface of the outer surface of the bearing metal and the surrounding air.

Continuity of the temperature and the heat flux is assumed also at the interface of the journal surface and the oil film. It is necessary to solve the energy equation of the oil film and the heat conduction equation within the journal simultaneously; however, it is difficult to solve the heat conduction equation within the journal because it is rotating and it is difficult to establish boundary conditions because the part of the journal extending outside the bearing may contact seals and other things and so its form is often complicated. Therefore, the following simplified method is used to avoid the solution of the heat conduction equation in the journal.

Let a part of the total amount of heat generated in the oil film $Q_j = \eta_{fj} Q_G$ be the heat dissipated to the surroundings through the journal and the shaft, where Q_G is the total amount of heat generated and η_{fj} is a coefficient. Then the integration of the local heat flux from the oil to the journal over the whole area of their interface must be equal to $\eta_{fj}Q_G$. Let the surface temperature of the journal be uniform in the circumferential direction because of the shaft rotation. In the axial direction, give the initial temperature distribution along a geometrical generator of the surface, then obtain the temperature distribution in the oil film by iterative calculations until the above condition has been satisfied. The details within the journal are not considered here. If the temperature distribution in the oil film is determined, the surface temperature of the journal will also be determined from the continuity of the temperature at the interface. Coefficient η_{fj} can be determined experimentally, as explained later.

As for the heat flow, it is assumed as shown schematically in Fig. 8.14 that, of the total heat Q_G generated in the lubricant film, Q_c, Q_b, and $Q_j = \eta_{fj}Q_G$ are carried away from the lubricant film by the convection of the oil leaking from the bearing end, by the conduction to the bearing metal, and by the conduction to the journal, respectively, and the rest, Q_r, is brought to the oil supply position by the recirculating oil.

The flow rate of the oil immediately downstream of the oil groove, q_{st}, can be written as a sum of part of the supply oil q_{in} and part of the recirculated oil q_r as follows:

$$q_{st} = \eta_{in} q_{in} + \eta_{mix} q_r \tag{8.74}$$

where η_{in} is the ratio of the supply oil actually involved in lubrication to the supply oil and η_{mix} is the ratio of the recirculating oil that is mixed with the new oil at the oil supply position (mixing ratio). The following equation is obtained from Eq. 8.74, the specific heat of the oil being assumed constant:

$$q_{st}T_{st} = \eta_{in}q_{in}T_{in} + \eta_{mix}q_r T_r \tag{8.75}$$

where T_{st}, T_{in}, and T_r are the temperature of the oil immediately downstream of the oil groove, the new oil, and the recirculating oil, respectively. If $\eta_{in}q_{in}$ is eliminated by using Eq. 8.74, the temperature immediately downstream of the oil groove T_{st} is obtained as:

$$T_{st} = T_{in} + \eta_{mix}(T_r - T_{in})q_r/q_{st} \tag{8.76}$$

8.6.3 Comparison of Calculated Results and Experiments

Specifications of the bearing and the lubricating oil used for calculations and experiments are shown in Table 8.2. Further, the temperature dependence of the oil viscosity is as shown in Fig. 8.15. Equation 8.72 is fitted to the curve in the figure in the practical range of 30° – 80°C. The viscosity of the oil at 40°C is shown in Table 8.2.

For the measurement of the temperature distribution, 144 copper–constantan thermocouples were embedded in the bearing metal of each bearing. Also, two gap sensors of the eddy current type were used for each journal to measure its displacement.

8 Heat Generation and Temperature Rise

Table 8.2. Specifications of the circular journal bearing and lubricant oil

Inner radius of metal	$D = 100$ mm	Clearance ratio c/R
Length of metal	$L = 70$ mm	$= 0.000787$ (bearing #1)
Length of oil groove	60 mm (axial)	$= 0.00101$ (bearing #2)
Width of oil groove	8.7 mm (circum.)	$= 0.00157$ (bearing #3)
Oil supply pressure	$P_{in} = 98$ kPa	$= 0.00222$ (bearing #4)
Oil supply temperature	$T_{in} = 40°C$	Bearing load $P = 490 - 7850$ N
Ambient temperature	$T_a = 15° - 30°C$	Rotating speed $N = 250 - 3500$ rpm

Oil properties at 40°C	Viscosity μ (mPa·s)	Thermal conductivity k_o [W/(m °K)]	Specific heat c_p [kJ/(kg °K)]	Density ρ (kg/m³)
Transformer oil	7.36	0.140	1.97	862
Turbine oil (#90)	19.2	0.131	1.95	859
Turbine oil (#140)	46.9	0.130	1.94	865

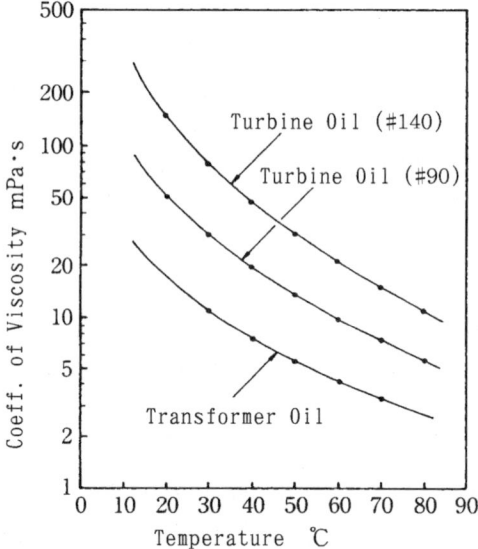

Fig. 8.15. Temperature dependence of the oil viscosity [22]

The values of η_{fj} and η_{mix} described above depend on the structure of the bearing, and must be determined experimentally. It is reported in the present case [17] [22] that approximately $\eta_{fj} = 0.04$, although there is some variation depending on load capacity, and approximately $\eta_{mix} = 0.5$, although it somewhat depends on the oil supply pressure. These values will be used hereafter.

Figure 8.16 shows comparisons of the experimental and theoretical temperature distribution at the middle cross section of the bearing metal for two different rotating speeds (some other conditions are slightly different, but are not significant). The operating conditions in Figs. 8.16a,b are, respectively, rotating speed $N = 1751$ rpm, 2502 rpm; load on the journal $P = 5.68$ kN, 5.61 kN; oil supply pressure $P_{in} = 98$ kPa, 98 kPa; oil supply temperature $T_{in} = 40.1°C$, $40.0°C$; and ambient temperature $T_a = 27.0°C$ and $29.2°C$. The bearing used was #3 bearing in both cases and the lubricating oil was transformer oil and #90 turbine oil. The eccentricity ratios were $\kappa = 0.8$ and 0.7, and the attitude angle was $\theta = 37°$ and $43°$, respectively.

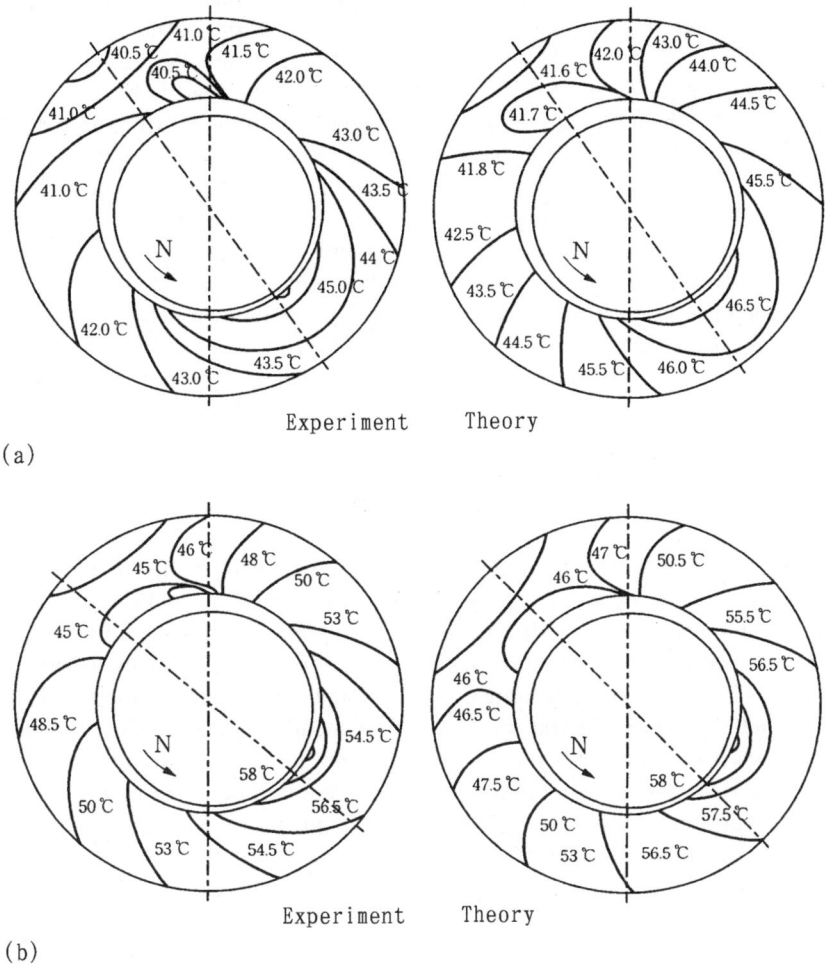

Fig. 8.16a,b. Temperature distribution in the middle section of the bearing metal [22]. **a** journal speed $N = 1751$ rpm, eccentricity ratio $\kappa = 0.8$, and the attitude angle $\theta = 37°$. **b** $N = 2502$ rpm, $\kappa = 0.7$, $\theta = 43°$

Experimental isotherms and theoretical isotherms are in good agreement in both Figs. 8.16a,b. The highest temperature in the bearing metal appears slightly downstream of the minimum film thickness point in Fig. 8.16a, and a little further downstream in Fig. 8.16b on the lubricating surface of the metal. The lowest temperature appears near the oil groove in both Figs. 8.16a,b.

Figure 8.17 shows calculated temperature distributions at the middle cross section of the oil film for peripheral velocities of $U = 10$ m/s and $U = 20$ m/s. The upper side of the rectangle corresponds to the journal surface and the lower side to the metal surface. The film thickness is very much exaggerated. The operating conditions, other than those shown in the figure, are the load on journal $P_1 = 5.68$ kN, the oil supply pressure $p_{in} = 98$ kPa, the oil supply temperature $T_{in} = 39.9° - 40.1°$C, ambient temperature $T_a = 25.0° - 27.8°$C, and the bearing used is bearing #3. The figures show that the temperature distribution in the oil film is far from uniform.

Comparison of Figs. 8.17a,b shows that the highest temperature in the oil film is higher in the case of higher peripheral speed, the highest temperatures being 54.7°C and 72.3°C. Further, while the highest temperature (a black dot shows the position) in the oil film appears a little upstream of the minimum film thickness position (the arrow from the letter h_{min}) in Fig. 8.17a, its position moves in the direction of rotation with the increase in peripheral speed, and the highest temperature appears a little downstream of the minimum film thickness position in Fig. 8.17b. In other words the locus of the highest temperature moves further than the minimum film thickness position does. The position of the highest temperature in the oil film agrees approximately with the position of the highest temperature of the bearing metal surface, and their values are similar.

Figure 8.18 shows the circumferential temperature distribution on the inner surface of the bearing metal (middle of the width) for four values of bearing clearance ratio c/R. The operating conditions are: the rotating speed of the shaft $N = 2250$ rpm, the bearing load $P_1 = 3.92$ kN, the oil supply pressure $p_{in} = 98$ kPa, the oil supply temperature $T_{in} = 39.9° - 40.2°$C, the ambient temperature $T_a = 17.6° - 22.5°$C, and the lubricating oil is #90 turbine oil. The theory roughly agrees with the experiment. For small bearing clearance, however, the experimental oil temperature is higher than the theory in the startup region of the metal temperature since the mixing coefficient η_{mix} was higher than the assumed value. In the figure, the bearing metal temperature rises considerably over the whole circumference with reduction of the clearance ratio, and the position of the highest temperature (the black dot) moves in the opposite direction to the rotation of the journal. Further, the eccentricity ratio (shown in parentheses) increases with decrease in the clearance ratio. This is due to the oil temperature rise and is an interesting phenomenon. Corresponding to this, the position of the minimum clearance (shown by an arrow) also moves in the opposite direction to the journal rotation. It is interesting that the position of the highest temperature moves more with decrease in the clearance ratio than the position of the minimum clearance does, and the position of maximum temperature has moved from the downstream side to the upstream side of the minimum thickness position.

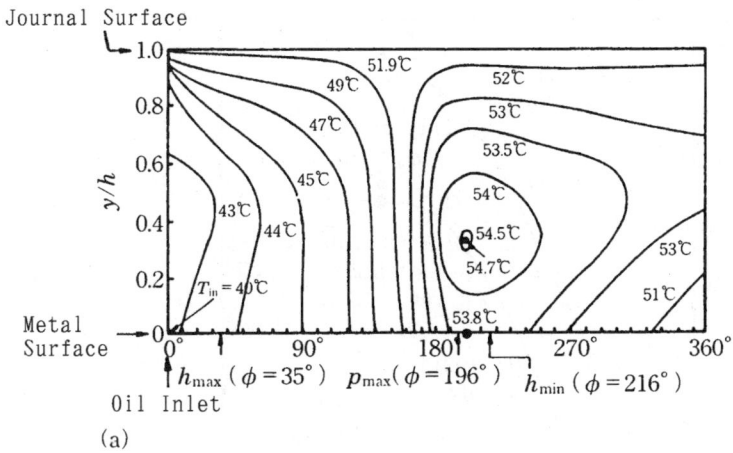

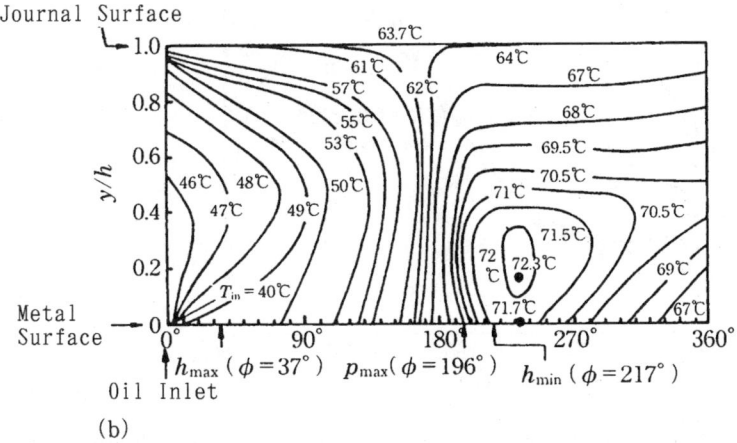

Fig. 8.17a,b. Temperature distribution at the middle cross section of the oil film (theory) [22]. **a** $U = 10$ m/s, **b** $U = 20$ m/s. The *black dots* show the highest temperatures of the oil film and the bearing metal. $\kappa = 0.8$, $c/R = 0.00157$, and the oil used was transformer oil

References

1. H. Lamb, "Hydrodynamics", *Dover*, New York, 1945, Sixth Edition, pp. 571-575.
2. R.B. Bird, W.E. Stewart, E.N. Lightfoot, "Transport Phenomena", *John Wiley & Sons, Inc.*, New York, 1960, Chapter 10.
3. D. Dowson, "A Generalized Reynolds Equation for Fluid Film Lubrication", International Journal of Mechanical Sciences, *Pergamon Press*, Vol. 4, 1962, pp. 159-170.
4. D. Dowson, J.D. Hudson, B. Hunter, C.N. March, "An Experimental Investigation of the Thermal Equibrium of Steadily Loaded Journal Bearings", *Proc. I. Mech. E.*, Vol. 181, Part 3B, 1966-1967, pp. 70-80.

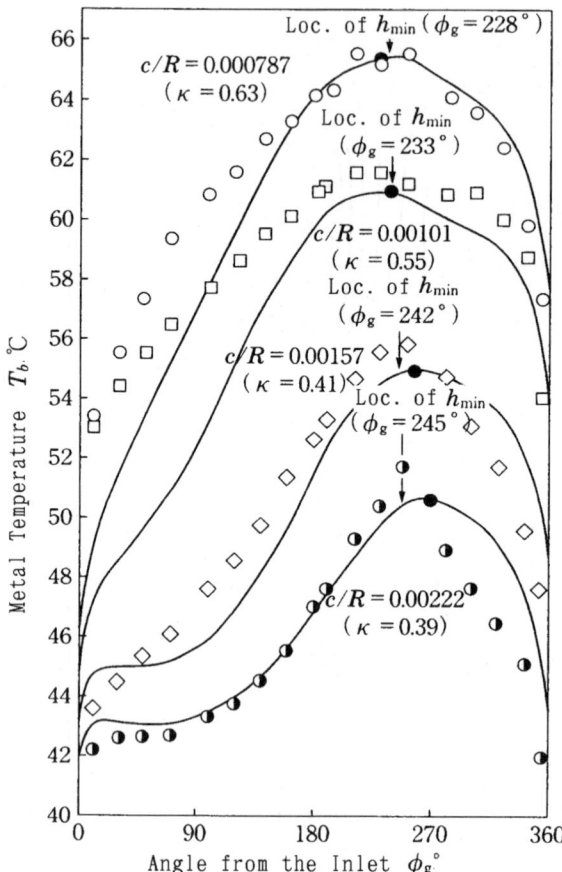

Fig. 8.18. Temperature distribution on the lubricating surface of the bearing metal (the influence of clearance ratio, theory and experiment) [22]. The *line* shows the theoretical values and the *symbols* show the experimental results for the c/R ratios indicated. The *black dots* show the locus of the maximum temperature

5. D. Dowson, C.N. March, "A Thermohydrodynamic Analysis of Journal Bearings", *Proc. I. Mech. E.*, Vol. 181, Part 3O, 1966-1967, pp. 117 - 126.
6. R.G. Woolacott, W.L. Cooke, "Thermal Aspects of Hydrodynamic Journal Bearing Performance at High Speeds", *Proc. I. Mech. E.*, Vol. 181, Part 3O, 1966-1967, pp. 127 - 135.
7. H. McCallion, F. Yousif, T. Lloyd, "The Analysis of Thermal Effects in a Full Journal Bearing", *Trans. ASME, Journal of Lubrication Technology*, Vol. 92, No. 4, 1970, pp. 578 - 587.
8. P. Fowles, "A Simpler Form of the General Reynolds Equation", *Trans. ASME, Journal of Lubrication Technology*, October 1970, Vol. 92, pp. 661 - 662.

9. H.A. Ezzat, S.M. Rhode, "A Study of the Thermohydrodynamic Performance of Finite Slider Bearings", *Trans. ASME, Journal of Lubrication Technology*, Vol. 95, No. 3, July 1973, pp. 298-307.
10. A.K. Tieu, "A Numerical Simulation of Finite-Width Thrust Bearings, Taking into Account Viscosity Variation with Temperature and Pressure", *Journal of Mechanical Enginneering Science*, Vol. 17, No. 1, 1975, pp. 1-10.
11. C. Ettles, "The Development of a Generalized Computer Analysis for Sector Shaped Tilting Pad Thrust Bearings", *Trans. ASLE*, Vol. 19, No. 2, April 1976, pp. 153-163.
12. T. Suganami, T. Masuda, A. Yamamoto and K. Sano, "The Effect of Varying Viscosity on the Performance of Journal Bearings" (in Japanese), *Journal of Japan Society of Lubrication Engineers*, Vol. 21, No. 8, August 1976, pp. 519-526.
13. T. Suganami, A.Z. Szeri, "A Thermohydrodynamic Analysis of Journal Bearings", *Trans. ASME, Journal of Lubrication Technology*, Vol. 101, No. 1, 1979, pp. 21-27.
14. R. Boncompain, J. Frene, "Thermohydrodynamic Analysis of a Finite Journal Bearings Static and Dynamic Characteristics", *Proc. I. Mech. E.*, Paper I(iii), 1980, pp. 33-44.
15. O. Pinkus, J.W. Lund, "Centrifugal Effects in Thrust Bearings and Seals under Laminar Conditions", *Trans. ASME, Journal of Lubrication Technology*, Vol. 103, No. 1, January 1981, pp. 126-136.
16. K.W. Kim, M. Tanaka, Y. Hori, "A Three-Dimensional Analysis of Thermohydrodynamic Performance of Sector-Shaped, Tilting-Pad Thrust Bearings", *Trans. ASME, Journal of Lubrication Technology*, Vol. 105, July 1983, pp. 406-413.
17. J. Mitsui, Y. Hori, M. Tanaka, "Thermodynamic Analysis of Cooling Effect of Supply Oil in Circular Journal Bearings", *Trans. ASME, Journal of Lubrication Technology*, Vol. 105, July 1983, pp. 414-421.
18. J. Ferron, J. Frene, R. Boncompain, "A Study of the Thermohydrodynamic Performance of a Plain Journal Bearing Comparison Between Theory and Experiments", *Trans. ASME, Journal of Lubrication Technology*, Vol. 105, No. 3, 1983, pp. 422-428.
19. M. Tanaka, Y. Hori and R. Ebinuma, "Measurement of the Film Thickness and Temperature Profiles in a Tilting Pad Thrust Bearing", *Proceedings JSLE International Tribology Conference*, July 8-10, 1985, Tokyo, Japan, pp. 553-558
20. K.W. Kim, M. Tanaka, Y. Hori, "Pad Attitude and THD Performance of Tilting Pad Thrust Bearings" (in Japanese), *Journal of Japan Society of Lubrication Engineers*, Vol. 31, No. 10, October 1986, pp. 741-748.
21. K.W. Kim, M. Tanaka, Y. Hori, "A Study on the Thermohydrodynamic Lubrication of Tilting Pad Thrust Bearing - The Effect of Inertia Force on the Bearing Performance -" (in Japanese), *Journal of Japan Society of Lubrication Engineers*, Vol. 31, No. 10, October 1986, pp. 749-755.
22. J. Mitsui, Y. Hori, M. Tanaka, "An Experimental Investigation on the Temperature Distribution in Circular Journal Bearings", *Trans. ASME, Journal of Lubrication Technology*, Vol. 108, October 1986, pp. 621-627.
23. K. W. Kim, M. Tanaka and Y. Hori, "An Experimental Study on the Thermohydrodynamic Lubrication of Tilting Pad Thrust Bearings" (in Japanese), *Journal of Japanese Society of Tribologists*, Vol. 40, No. 1, January 1995, pp. 70-77.

9

Turbulent Lubrication

In Reynolds' theory of lubrication, the flow in a lubricant film is assumed to be laminar. In large, high speed bearings in recent years, however, the flow is often turbulent. In this case, the shear resistance and heat generation in the fluid film increases markedly. And what is worse, the flow rate of the oil will decrease. These are big problems for bearings. On turbulence in bearings, since Wilcock's experimental work (1950) [3] and Constantinescu's theoretical contribution (1959) [7], many studies have been carried out [9] [11] [12] [15] [16] [17] [26] [27]. While most analyses in the past were based on Prandtl's mixing length hypothesis, more general analyses based on the k-ε model will also be described in this chapter.

Turbulence is a big problem in a fluid seal also. Although a fluid seal is similar to a journal bearing in form, it differs in that the axial pressure gradient and hence the axial flow velocity is large in a fluid seal. In a fluid seal, both high speed rotation and steep pressure gradients cause turbulence. In this chapter, fluid seals are also considered.

In a thin fluid film, it is known that the transition from laminar flow to turbulent flow takes place when the bearing Reynolds' number Re reaches approximately 1000, where Re is defined as follows with circumferential speed U, film thickness h ($= c$), and kinetic viscosity v:

$$Re = \frac{Uc}{v} \approx 1000 \tag{9.1}$$

If a large bearing, 600 mm in diameter and 0.6 mm in radial clearance, for a steam turbogenerator is considered, and if the kinetic viscosity of the oil used is 25 cSt, the transitional speed of the shaft at which the transition from laminar to turbulent flow takes place in the fluid film is calculated to be 1326 rpm. Since the rated speed of generators is usually 3000 or 3600 rpm, the flow in the fluid film becomes turbulent very easily.

9.1 Time-Average Equation of Motion and the Reynolds' Stress

A turbulent shear flow, as shown in Fig. 9.1, is considered. An average flow is assumed to be parallel to the x axis. In turbulent flow, eddies (the blobs of fluid with some definitive character) of the fluid of various sizes go back and forth violently between the layers of different velocities and thus exchange momentum. Shear resistance arises as a result of this, somewhat similar to the way viscous resistance of a gas arises as a result of exchange of momentum by molecular motion. In a turbulent fluid, however, the exchange of momentum by the eddies of fluid is very large, which causes very large shear resistance in a turbulent fluid. This phenomenon will be considered below [10] [14].

While the shear resistance of a turbulent fluid is the sum of the resistances due to momentum exchange and that due to fluid viscosity, the latter is usually small and can be disregarded compared with the former. In the neighborhood of a solid wall, however, the momentum exchange is small and the contribution of viscosity becomes significant.

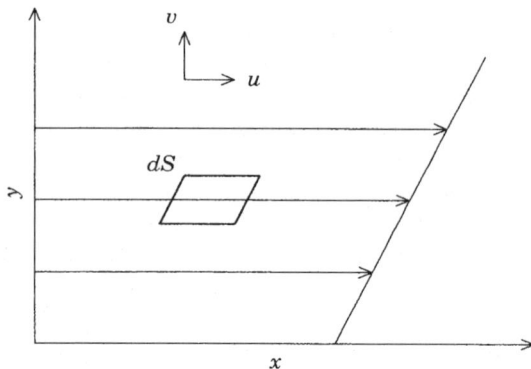

Fig. 9.1. Reynolds' stress

The turbulent shear stress due to the exchange of momentum by eddies is obtained as follows. Although the turbulent shear stress is an unsteady quantity in nature, only its time average will be considered here because it satisfies most practical needs.

In the case of turbulent flow, the components of velocity u and v and the pressure p of a small volume of fluid can be expressed as the sum of their time average (steady part) and fluctuations (unsteady part) as follows:

$$u = \bar{u} + u', \quad v = \bar{v} + v', \quad p = \bar{p} + p' \tag{9.2}$$

where ($\bar{}$) shows the time average or the steady part, and ($'$) indicates the unsteady part. Since the time average of the unsteady part is zero, and the time average flow is

9.1 Time-Average Equation of Motion and the Reynolds' Stress

assumed to be parallel to the x axis, the following relations will be obtained:

$$\overline{u'} = \overline{v'} = \overline{p'} = 0, \qquad \overline{v} = 0 \tag{9.3}$$

Now, consider a small area dS in the fluid. dS is perpendicular to the y axis as shown in Fig. 9.1. The volume of the fluid that passes the area dS in the positive direction of y during a time interval dt is $v \cdot dS \cdot dt$. The x component of the momentum carried by this volume of fluid is $\rho uv \cdot dS \cdot dt$, ρ being the density of the fluid. Thus, the flow of momentum per unit area and unit time is equal to ρuv. This gives the turbulent shear stress, if the sign is changed:

$$\tau_t = -\rho \, \overline{uv} \tag{9.4}$$

The negative sign in the above equation comes from the customary sign of the shear stress.

Now, consider the time average of the turbulent shear stress τ_t. It can be written as follows by using Eqs. 9.2 and 9.3:

$$\tau_t = -\rho \, \overline{(\overline{u} + u')v'} = -\rho \, \overline{u'v'} \tag{9.5}$$

A horizontal line over each symbol indicates the time average. Thus, the turbulent shear stress is given by the correlation of the unsteady parts of the velocity of the fluid. This idea was proposed by Reynolds and $-\rho \, \overline{u'v'}$ in the above equation is called the **Reynolds' stress**.

Let us consider the sign of τ_t. In the case of a shear flow where $d\overline{u}/dy > 0$, it is known that, in practice, if $v' > 0$ then $u' < 0$ and if $v' < 0$ then $u' > 0$, respectively, with a high probability. Therefore, the probability that $u'v' < 0$ is very high, and so $\overline{u'v'}$ becomes negative. Therefore, τ_t is positive when $d\overline{u}/dy > 0$.

Considering the time average of the Navier–Stokes equation leads to a more general derivation of the Reynolds' stress. First, write down the Navier–Stokes equation in the x direction and in the y direction as follows, where τ_{ij} represents a stress component acting on plane i in direction j:

$$\rho \left(\frac{\partial u}{\partial t} + u \frac{\partial u}{\partial x} + v \frac{\partial u}{\partial y} \right) = -\frac{\partial p}{\partial x} + \frac{\partial \tau_{xx}}{\partial x} + \frac{\partial \tau_{yx}}{\partial y} \tag{9.6}$$

$$\rho \left(\frac{\partial v}{\partial t} + u \frac{\partial v}{\partial x} + v \frac{\partial v}{\partial y} \right) = -\frac{\partial p}{\partial y} + \frac{\partial \tau_{xy}}{\partial x} + \frac{\partial \tau_{yy}}{\partial y} \tag{9.7}$$

Next, multiply the continuity equation for an incompressible fluid by ρ and u to give the following equation:

$$\rho \left(u \frac{\partial u}{\partial x} + u \frac{\partial v}{\partial y} \right) = 0$$

By using this relation, Eq. 9.6 in the x direction is rewritten as:

$$\rho \left(\frac{\partial u}{\partial t} + \frac{\partial (uu)}{\partial x} + \frac{\partial (uv)}{\partial y} \right) = -\frac{\partial p}{\partial x} + \frac{\partial \tau_{xx}}{\partial x} + \frac{\partial \tau_{yx}}{\partial y}$$

9 Turbulent Lubrication

Considering the time average of the above equation and using the relations $u = \bar{u}+u'$, $v = \bar{v}+v'$, $p = \bar{p}+p'$, $\overline{uu} = \overline{u}\overline{u}+\overline{u'u'}$ and $\overline{uv} = \overline{u}\overline{v}+\overline{u'v'}$ yields the following equation:

$$\rho\left(\frac{\partial \bar{u}}{\partial t} + \frac{\partial(\bar{u}\bar{u})}{\partial x} + \frac{\partial(\bar{u}\bar{v})}{\partial y}\right) = -\frac{\partial \bar{p}}{\partial x} + \frac{\partial}{\partial x}\left(\tau_{xx} - \rho\overline{u'u'}\right) + \frac{\partial}{\partial y}\left(\tau_{yx} - \rho\overline{u'v'}\right)$$

Let us return the left-hand side of this equation back to that of Eq. 9.6 with the help of

$$\rho\left(\bar{u}\frac{\partial \bar{u}}{\partial x} + \bar{u}\frac{\partial \bar{v}}{\partial y}\right) = 0$$

which is obtained from the continuity equation $\partial \bar{u}/\partial x + \partial \bar{v}/\partial y = 0$, giving the following equation:

$$\rho\left(\frac{\partial \bar{u}}{\partial t} + \bar{u}\frac{\partial \bar{u}}{\partial x} + \bar{v}\frac{\partial \bar{u}}{\partial y}\right) = -\frac{\partial \bar{p}}{\partial x} + \frac{\partial}{\partial x}\left(\tau_{xx} - \rho\overline{u'u'}\right) + \frac{\partial}{\partial y}\left(\tau_{yx} - \rho\overline{u'v'}\right)$$

This is the time average of the Navier–Stokes equation, i.e., a **time-average equation of motion** of the steady part of a turbulent flow (time-average flow). If this is compared with the Navier–Stokes equation (Eq. 9.6), it will be noticed that two new terms $-\rho\overline{u'u'}$ and $-\rho\overline{u'v'}$ have appeared on the right-hand side. These are the Reynolds' stresses (Reynolds 1895).

A similar equation can also be obtained in the y direction.

The time-average equations in the x and y directions are mentioned together below, where the overbars indicating the steady parts are omitted for simplicity:

$$\rho\left(\frac{\partial u}{\partial t} + u\frac{\partial u}{\partial x} + v\frac{\partial u}{\partial y}\right) = -\frac{\partial p}{\partial x} + \frac{\partial}{\partial x}\left(\tau_{xx} - \rho\overline{u'u'}\right) + \frac{\partial}{\partial y}\left(\tau_{yx} - \rho\overline{u'v'}\right) \quad (9.8)$$

$$\rho\left(\frac{\partial v}{\partial t} + u\frac{\partial v}{\partial x} + v\frac{\partial v}{\partial y}\right) = -\frac{\partial p}{\partial y} + \frac{\partial}{\partial x}\left(\tau_{xy} - \rho\overline{v'u'}\right) + \frac{\partial}{\partial y}\left(\tau_{yy} - \rho\overline{v'v'}\right) \quad (9.9)$$

Thus, the time-average equations of motion of a turbulent flow include Reynolds' stresses, namely, the terms of correlation of the fluctuations in the velocity in the parentheses of the right-hand side of the equations, and, in the case of the above equations, they are the four terms shown below. Because of symmetry, however, only three of them are different from each other.

$$\begin{pmatrix} -\rho\overline{u'u'} & -\rho\overline{u'v'} \\ -\rho\overline{v'u'} & -\rho\overline{v'v'} \end{pmatrix}$$

Of these Reynolds' stresses, the normal stress $-\rho\overline{u'u'}$ and $-\rho\overline{v'v'}$ are apparent pressures, and their influence is usually negligible. Of great importance is the shear stress $-\rho\overline{u'v'}$ and this coincides with Eq. 9.5.

Although Eqs. 9.8 and 9.9 are called Reynolds' equation in many books on turbulence, this name is not used in this book to avoid confusion with the previously used Reynolds' equation, the basic equation of lubrication.

9.2 Turbulent Flow Model

The time-average of turbulent flow can be obtained from simultaneous solutions of Eqs. 9.8 and 9.9. However, since the fluctuations in the velocity are unknown, Reynolds' stress $\tau_t = -\rho \overline{u'v'}$ cannot be calculated. Therefore, something additional is necessary to solve Eqs. 9.8 and 9.9.

If Eqs. 9.6 and 9.7 (the Navier–Stokes equation) are used together with Eqs. 9.8 and 9.9, the formula for the Reynolds' stress can be derived. However, new unknown quantities such as correlations of the third order of fluctuations and correlations including fluctuations of pressure appear in the formula, and if similar operations are repeated to obtain them, new unknown quantities will appear each time, and the system of equations will never close. Therefore, to solve Eqs. 9.8 and 9.9, certain assumptions must be made to reduce the number of unknown quantities so that the system of equations will close. The assumptions on the structure of turbulence for this purpose form the turbulence model.

Typical turbulence models include (1) the mixing length model and (2) the k-ε model (k = turbulent flow energy, ε = turbulent flow loss). When the pressure gradient is not very large (when the eccentricity ratio is small in the case of bearings), the mixing length model will suffice; when the pressure gradient is large and reverse flow arises in the fluid film (when the eccentricity ratio is large in the case of bearings), since the pressure gradient affects the structure of turbulence, it is necessary to use a more fundamental model, the k-ε model.

9.2.1 Mixing Length Model

It is assumed that an eddy that is performing violent irregular motions in a turbulent flow travels by a certain distance and is mixed with the fluid at the end of the travel, resulting in the exchange of momentum. The average distance of motion is called the **mixing length** and is represented by l. The size of fluctuations in the velocity in the x direction $|u'|$ will be of the order of $l\,|du/dy|$. The size of fluctuations in the velocity in the y direction $|v'|$ will be of the same order of magnitude as $|u'|$. This is because u' and v' are attributable to the motion of the same eddy, i.e.,

$$|u'| \approx |v'| \approx l \left|\frac{du}{dy}\right| \tag{9.10}$$

When $d\overline{u}/dy > 0$, since $\overline{u'v'}$ is negative as mentioned above, the following equation is obtained, by using the above equation:

$$\overline{u'v'} \approx -|u'||v'| \approx -l^2 \left(\frac{du}{dy}\right)^2 \tag{9.11}$$

Therefore, Reynolds' stress (turbulent flow shearing stress) $\tau_t = -\rho \overline{u'v'}$ can be written as follows:

$$\tau_t = -\rho \overline{u'v'} = \rho\, l^2 \left(\frac{du}{dy}\right)^2 \tag{9.12}$$

Or, to take the sign into consideration, it is written as follows with the symbol of absolute value:

$$\tau_t = -\rho \overline{u'v'} = \rho l^2 \left|\frac{du}{dy}\right| \frac{du}{dy} \quad (9.13)$$

The approach described above is called **Prandtl's mixing length model** (Prandtl 1925).

If τ_t is expressed, after a viscous stress, in the form of (coefficient) × (gradient of average velocity of turbulent flow), Eq. (9.13) will be:

$$\tau_t = -\rho \overline{u'v'} = \mu_t \frac{du}{dy} \quad (9.14)$$

where μ_t is:

$$\mu_t = \rho l^2 \left|\frac{du}{dy}\right| \quad (9.15)$$

Although μ_t is called the **turbulent viscosity coefficient**, it is clearly a quantity that depends on the internal structure of the turbulence, and is not a material constant.

The mixing length l in the above theory is an unknown quantity depending on the distance from the wall, the velocity gradient, and so on, and is given by an empirical formula. Among various formulae proposed, the simplest one is to assume that the mixing length l is proportional to the distance from the wall, i.e.,

$$l = \kappa_k y \quad (9.16)$$

where y is the distance from the wall and κ_k is a proportionality constant called Kármán's constant.

The velocity distribution in the turbulent boundary layer in this case is calculated as follows. Let the surface shear stress be τ_w and assume that the shear stress is constant in the neighborhood of the wall, i.e., $\tau_t = \tau_w$ = constant. Then, Eq. 9.12 can be written as:

$$\frac{\tau_w}{\rho} = (\kappa_k y)^2 \left(\frac{du}{dy}\right)^2 \quad (9.17)$$

This can be rewritten further as:

$$\frac{du}{dy} = \frac{u^*}{\kappa_k y} \quad (9.18)$$

where $u^* = \sqrt{\tau_w/\rho}$ is a quantity with the dimension of velocity and is called the **friction velocity**. Integrating Eq. 9.18 gives the velocity distribution as follows:

$$u = \frac{u^*}{\kappa_k} \ln y + C \quad (9.19)$$

This is called the **logarithmic law** of velocity distribution.

The following formula is a modification of Eq. 9.16 that takes the anisotropy of eddies immediately near the wall into consideration:

$$l = \kappa_k y \left[1 - \exp(-y/A)\right] \quad (9.20)$$

This is called van Driest's formula [5].

9.2.2 k-ε Model

The mixing length l in the mixing length model is given by an empirical formula, the constants of which change with pressure gradient. The constants are usually determined experimentally under relatively low pressure gradients, therefore their use is questionable in the case of steep pressure gradients (when the eccentricity ratio is large in a bearing). A more reasonable turbulent model is the k-ε model in which k is the turbulent energy and ε is the turbulent loss [20] [38] [39] [40]. Although experimental constants are required in this case also, they are almost universal constants and hardly change with the pressure gradient; k-ε models are excellent in this respect.

The k-ε models include high-Reynolds' number models (standard models) and low-Reynolds' number models. In the case of a lubricating film, especially in the neighborhood of the wall surface, the low-Reynolds' number model is suitable, because in these cases the turbulent Reynolds' number $R_t = k^2/(\varepsilon v)$ is comparatively low. The low-Reynolds' number k-ε model, which is applicable up to the wall surface, was proposed by Jones and Launder [21] [22] as follows:

If the turbulent energy k and the turbulent loss ε are defined as

$$k = \frac{1}{2}\overline{u_i' u_i'}, \quad \varepsilon = v \overline{\frac{\partial u_i'}{\partial x_j} \frac{\partial u_i'}{\partial x_j}}, \quad (9.21)$$

then the transport equation of k and that of ε are written as follows, using the turbulent Reynolds number $R_t = k^2/(\varepsilon v)$:

$$\frac{Dk}{Dt} = \frac{\partial}{\partial y}\left[\left(v + \frac{v_t}{\sigma_k}\right)\frac{\partial k}{\partial y}\right] + v_t \left(\frac{\partial u}{\partial y}\right)^2 - \varepsilon - 2v\left(\frac{\partial k^{1/2}}{\partial y}\right)^2 \quad (9.22)$$

$$\frac{D\varepsilon}{Dt} = \frac{\partial}{\partial y}\left[\left(v + \frac{v_t}{\sigma_\varepsilon}\right)\frac{\partial \varepsilon}{\partial y}\right] + C_{\varepsilon 1} v_t \left(\frac{\partial u}{\partial y}\right)^2 \frac{\varepsilon}{k}$$

$$- C_{\varepsilon 2}\left[1 - 0.3\exp(-R_t^2)\right]\frac{\varepsilon^2}{k} + 2v v_t \left(\frac{\partial^2 u}{\partial y^2}\right)^2 \quad (9.23)$$

where σ_k, σ_ε, $C_{\varepsilon 1}$, $C_{\varepsilon 2}$ and C_μ are experimental constants, which are almost universal and hardly dependent on pressure gradients, as stated before. This is the most advantageous point of the k-ε models.

Further, the turbulent viscosity coefficient v_t is given as follows using k and ε:

$$v_t = C_\mu \frac{k^2}{\varepsilon} \quad (9.24)$$

In k-ε model analyses, generally speaking, the time-average momentum equation (Reynolds' stresses are included), the transport equation of turbulent energy k, and that of turbulent loss ε are solved simultaneously. Then, Reynolds' stress is given as follows by using ν_t:

$$-\rho \overline{u'v'} = \rho \nu_t \frac{\partial u}{\partial y} \qquad (9.25)$$

The system of equations is now closed.

9.3 Turbulent Lubrication Theory Using the Mixing Length Model

Turbulent hydrodynamic lubrication of a bearing and a fluid seal are considered here using a modified mixing length model.

9.3.1 Modified Mixing Length

The fluid film in a bearing or a fluid film seal is so thin that the influence of the viscous sublayer in the immediate neighborhood of a wall cannot be ignored even in the case of turbulent flow. To deal with turbulence systematically from a turbulent zone to the viscous sublayer by a mixing length model, a modified mixing length is introduced (Hori, Fukayama, et al. [33]).

If the shear stress is considered to be the sum of the viscous shear stress and the turbulent shear stress of Eq. 9.12, the total shear stress τ can be written as:

$$\tau = \mu \frac{\partial u}{\partial y} + \rho l^2 \left(\frac{\partial u}{\partial y}\right)^2 \qquad (9.26)$$

If this is nondimensionalized by using the friction velocity $u^* = \sqrt{\tau_w/\rho}$,

$$\tau^+ = \frac{\partial u^+}{\partial y^+} + l^{+2} \left(\frac{\partial u^+}{\partial y^+}\right)^2 \qquad (9.27)$$

will be obtained, where

$$\tau^+ = \frac{\tau}{\tau_w}, \quad u^+ = \frac{u}{u^*}, \quad l^+ = \frac{u^* l}{\nu}, \quad y^+ = \frac{u^* y}{\nu} \qquad (9.28)$$

To describe the mixing length, van Driest's formula is used. It can be written in a nondimensional form as follows:

$$l^+ = \kappa_k y^+ \left[1 - \exp(-y^+/A^+)\right] \qquad (9.29)$$

where $\kappa_k = 0.4$ and $A^+ = Au^*/\nu = 26$.

9.3 Turbulent Lubrication Theory Using the Mixing Length Model

To express Eq. 9.27 approximately in the form of Eq. 9.12 for the sake of convenience in calculation, a suitable modified mixing length is proposed. Let the wall shear stress be τ_w and assume that the shear stress τ_t is constant near the wall surface. Then, $\tau_t = \tau_w$ = constant, therefore $\tau^+ = 1$. If we introduce $\tau^+ = 1$ into Eq. 9.27 and solve it as a quadratic equation of $(\partial u^+/\partial y^+)$, then the following result will be obtained after some rearrangement:

$$\frac{\partial u^+}{\partial y^+} = \frac{2}{1 + \sqrt{1 + 4\,l^{+2}}} \tag{9.30}$$

Now let us consider Eqs. 9.29 and 9.30. In the immediate neighborhood of the wall surface, $l^+ \approx 0$ is obtained from Eq. 9.29 because y^+ is small enough. Therefore, $(\partial u^+/\partial y^+) = 1$ is obtained from Eq. 9.30 and thus $u^+ = y^+$. This expresses the velocity distribution in the viscous sublayer. In the region far enough from the wall surface, $\exp(-y^+/A^+) \approx 0$ because y^+ is very large, therefore $l^+ = \kappa_k y^+$ is obtained from Eq. 9.29. And since y^+ is large, $(\partial u^+/\partial y^+) \approx 1/(\kappa_k y^+)$ results from Eq. 9.30. Integrating this with respect to y^+ yields $u^+ = (1/\kappa_k) \ln y^+ +$ constant. This gives the velocity distribution in the logarithmic region. It turns out from the above discussion that the velocity distribution is well expressed by Eq. 9.30 not only in the viscous sublayer but also in the logarithmic region.

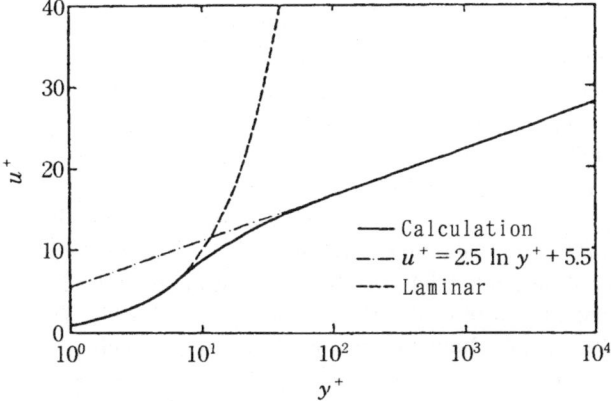

Fig. 9.2. Velocity distribution near the wall surface [33]

Then, introducing the modified mixing length

$$l_m^+ = \frac{1 + \sqrt{1 + 4\,l^{+2}}}{2} \tag{9.31}$$

and examining

$$\tau^+ = l_m^{+2}\left(\frac{\partial u^+}{\partial y^+}\right)^2 \tag{9.32}$$

reveals that this is of the same form as Eq. 9.12, and should apply in the region of $\tau^+ \approx 1$. Figure 9.2 shows the velocity distribution near the wall surface calculated using Eq. 9.32 with $\tau^+ = 1$. In the figure, the calculated result is in good agreement with the empirical formula $u^+ = y^+$ in the viscous sublayer and with the empirical formula $u^+ = 2.5 \ln y^+ + 5.5$ in the logarithmic region. This shows the validity of the modified mixing length l_m^+.

While a one-dimensional flow has been considered so far, the same modified mixing length is assumed to be applicable also to a two-dimensional flow.

Various formulae are proposed for the modified mixing length. For example, a modified mixing length of the form of $l_m = ay$ is proposed with the following coefficient of modification a (Aoki and Harada [17]):

$$a(y^+) = \frac{1}{y^+} + \frac{Ay^+}{1 + By^+} \tag{9.33}$$

Comparison with the known velocity distribution determines the constants in the formula as $A = 0.016$ and $B = 0.04$.

9.3.2 Turbulent Velocity Distribution Between Two Surfaces

A turbulent film in a bearing or a seal is different from the turbulence in a semi-infinite space in that the former is held between two surfaces. Turbulent flow in a narrow clearance between two surfaces, as shown in Fig. 9.3, is considered here. For a coordinate system, the x and z axes are taken on one of the surfaces and the y axis is taken in the thickness direction.

For simplicity, let us divide the clearance into two domains, domain 1 ($0 < y < y_m$) and domain 2 ($y_m < y < h$). It is assumed that the flow in domain 1 is influenced only by wall surface 1 and the flow in domain 2 is only influenced by wall surface 2. In other words, it is assumed that the flow velocity and the mixing length can be calculated independently from wall surfaces 1 and 2, respectively. These two quantities must be continuous at the boundary $y = y_m$ of the two domains.

When body forces and inertia forces can be ignored, the equation of balance of an element of incompressible fluid can be written as follows:

$$\frac{\partial \tau_x}{\partial y} = \frac{\partial p}{\partial x}, \quad \frac{\partial \tau_z}{\partial y} = \frac{\partial p}{\partial z} \tag{9.34}$$

Integrating these equations with respect to coordinates y_1 and y_2, which are taken from wall surface 1 and 2 in the film thickness direction, respectively, yields:

$$\tau_x = \tau_{x1} + \frac{\partial p}{\partial x} y_1 = \tau_{x2} - \frac{\partial p}{\partial x} y_2 \tag{9.35}$$

$$\tau_z = \tau_{z1} + \frac{\partial p}{\partial z} y_1 = \tau_{z2} - \frac{\partial p}{\partial z} y_2 \tag{9.36}$$

9.3 Turbulent Lubrication Theory Using the Mixing Length Model

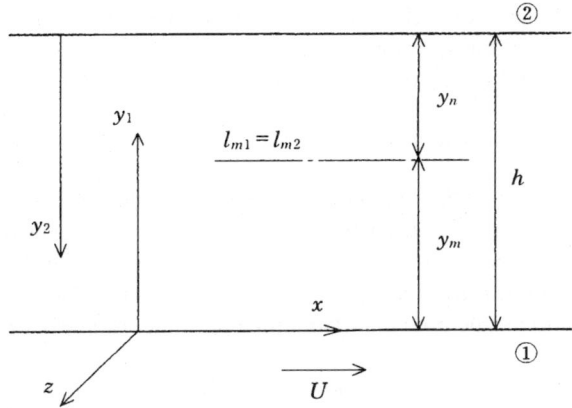

Fig. 9.3. Turbulent flow in clearance [33]

where τ_{x1}, τ_{x2} and τ_{z1}, τ_{z2} are shear stresses on the wall surfaces 1 and 2 in the x and z directions, respectively. Or, using the relation $y_1 + y_2 = h$, we have:

$$\tau_{x1} - \tau_{x2} = -\frac{\partial p}{\partial x} h \tag{9.37}$$

$$\tau_{z1} - \tau_{z2} = -\frac{\partial p}{\partial z} h \tag{9.38}$$

Additionally, the turbulent shear stress is given by Eq. 9.32, as explained before, using the modified mixing length. While this is a one-dimensional equation, the turbulent shear stresses in a two-dimensional case are assumed to be given as:

$$\tau_x = \rho \, l_m^2 \sqrt{\left(\frac{\partial u}{\partial y}\right)^2 + \left(\frac{\partial w}{\partial y}\right)^2} \cdot \frac{\partial u}{\partial y} \tag{9.39}$$

$$\tau_z = \rho \, l_m^2 \sqrt{\left(\frac{\partial u}{\partial y}\right)^2 + \left(\frac{\partial w}{\partial y}\right)^2} \cdot \frac{\partial w}{\partial y} \tag{9.40}$$

where l_m is the modified mixing length (dimensional). If Eqs. 9.39 and 9.40 are solved with respect to $\partial u/\partial y$ and $\partial w/\partial y$, the following equations are obtained, in which $|\tau| = \sqrt{\tau_x^2 + \tau_z^2}$:

$$\frac{\partial u}{\partial y} = \frac{\tau_x/\rho}{\sqrt{|\tau|/\rho} \cdot l_m} \tag{9.41}$$

$$\frac{\partial w}{\partial y} = \frac{\tau_z/\rho}{\sqrt{|\tau|/\rho} \cdot l_m} \tag{9.42}$$

In the case of a journal bearing, it can usually be assumed that $\tau_x \gg \tau_z$, and in this case Eqs. 9.41 and 9.42 can be simplified. For a fluid film seal, it is not always the case that $\tau_x \gg \tau_z$, and so such simplifications are not permitted and Eqs. 9.41 and 9.42 will be used without simplification.

The boundary conditions for the velocity on the wall surfaces are as follows:

$$u = U, \ w = 0 \text{ at } y = 0; \quad u = w = 0 \text{ at } y = h \tag{9.43}$$

Integrating Eqs. 9.41 and 9.42 under these boundary conditions from wall surfaces 1 and 2, respectively, will give the velocities u_1, u_2 and w_1, w_2. These velocities must be continuous at the boundary $y_1 = y_m$ or $y_2 = y_n$ $(= h - y_m)$ of domains 1 and 2 in the film. Mixing lengths l_{m1} and l_{m2} calculated from wall surfaces 1 and 2 must also be continuous at the same boundary from the continuity of shear stress, i.e.,

$$|u_1(y_m)| + |u_2(y_n)| = U \tag{9.44}$$

$$|w_1(y_m)| = |w_2(y_n)| \tag{9.45}$$

$$l_{m1}(y_m) = l_{m2}(y_n) \tag{9.46}$$

If the function of l_m is given explicitly, the five above equations, Eqs. 9.37, 9.38, 9.44, 9.45, and 9.46, can be regarded as the simultaneous equations for τ_{x1}, τ_{x2}, τ_{z1}, τ_{z2}, and y_m with $\partial p/\partial x$, $\partial p/\partial z$, and U as parameters. If the solutions are found, the velocity distribution can be determined by Eqs. 9.41 and 9.42.

9.3.3 Turbulent Reynolds' Equation

As coefficients for the flow characteristics of a turbulent fluid, **turbulent coefficients** G_x and G_z in the x and z directions, respectively, are defined as follows:

$$G_x = \frac{1}{B_x}\left(\frac{1}{h}\int_0^h \frac{u}{U}dy - \frac{1}{2}\right) \tag{9.47}$$

$$G_z = \frac{1}{B_z}\left(\frac{1}{h}\int_0^h \frac{w}{U}dy\right) \tag{9.48}$$

where B_x and B_z are nondimensional pressure gradients defined as:

$$B_x = -\frac{h^2}{\mu U}\frac{\partial p}{\partial x}, \quad B_z = -\frac{h^2}{\mu U}\frac{\partial p}{\partial z} \tag{9.49}$$

where U is the surface velocity. Also, u and w are flow velocities that can be calculated as shown in the previous section.

It can be seen from Eqs. 9.47 and 9.48 that the turbulent coefficient is given by the ratio of the flow rate to the corresponding pressure gradient; the larger the turbulent coefficient, the larger the flow rate of the turbulent fluid. The term $-1/2$ in the parenthesis of Eq. 9.47 indicates subtraction of the Couette flow from the flow rate. The turbulent coefficients G_x and G_z are nondimensional quantities.

Combining the definition of the turbulent coefficients G_x and G_z and the continuity equation

9.3 Turbulent Lubrication Theory Using the Mixing Length Model

$$\frac{\partial}{\partial x}\int_0^h u\,dy + \frac{\partial}{\partial z}\int_0^h w\,dy = 0 \tag{9.50}$$

yields the turbulent Reynolds' equation as follows:

$$\frac{\partial}{\partial x}\left(G_x \frac{h^3}{\mu}\frac{\partial p}{\partial x}\right) + \frac{\partial}{\partial z}\left(G_z \frac{h^3}{\mu}\frac{\partial p}{\partial z}\right) = \frac{U}{2}\frac{\partial h}{\partial x} \tag{9.51}$$

This equation is also written as follows, using $k_x = 1/G_x$ and $k_z = 1/G_z$:

$$\frac{\partial}{\partial x}\left(\frac{h^3}{k_x \mu}\frac{\partial p}{\partial x}\right) + \frac{\partial}{\partial z}\left(\frac{h^3}{k_z \mu}\frac{\partial p}{\partial z}\right) = \frac{U}{2}\frac{\partial h}{\partial x} \tag{9.52}$$

Large k_x or large k_z values indicate a large flow resistance.

9.3.4 Turbulent Coefficients of Fluid Film Seals

The turbulent coefficients can be obtained from Eqs. 9.47 and 9.48 in the preceding section. Figure 9.4a,b shows some examples of turbulent coefficients G_x and G_z thus obtained in the case of a fluid film seal. The horizontal axis shows local Reynolds' number $R_h = Uh/\nu$ based on the circumferential velocity of the journal. B_x and B_z are pressure gradients.

First, Fig. 9.4a,ba shows the calculated results in the case $B_x = B_z$. As seen in the figure, the turbulent coefficients depend greatly on the pressure gradient; for example, when the pressure gradient is large, the turbulent coefficients G_x and G_z are small. This means that if the pressure gradient becomes large, the turbulence becomes violent and the fluidity of the fluid decreases. Therefore, the flow rate does not necessarily increase in proportion to the pressure gradient. The turbulent coefficient depends greatly also on Reynolds' number $R_h = Uh/\nu$. The larger R_h becomes, the smaller the resulting turbulent coefficient, i.e., the flow rate will decrease if Reynolds' number increases, the pressure gradient being the same. Further, G_x and G_z are different for the same pressure gradient; this means that the fluidity in the x and y directions are different.

Figure 9.4a,bb shows the results in the case $B_x \neq B_z$. As in Fig. 9.4a,ba, the larger the pressure gradient or Reynolds' number, the smaller the turbulent coefficients. The difference in the turbulent coefficients of curves 1 and 3 corresponds to the difference in flow rates in the case that the main pressure flow (circumferential direction in 1 and axial direction in 3) and the shear flow (circumferential direction) are parallel and in the case that they are perpendicular to each other. Although the same is true of curves 2 and 4, the difference is very small in this case. This may be because the pressure flow is predominant and the influence of shear flow is relatively small.

The turbulent coefficients for a journal bearing (curve 4, Aoki and Harada [17]) are also shown in Fig. 9.4a,ba. It is generally accepted that the pressure gradient in a journal bearing is not very large and that the average values of both B_x and B_z are, roughly speaking, below 10. The turbulent coefficients in this case are actually in good agreement with those for $B_x = B_z = 10$. The turbulent coefficients in this case are expressed as follows:

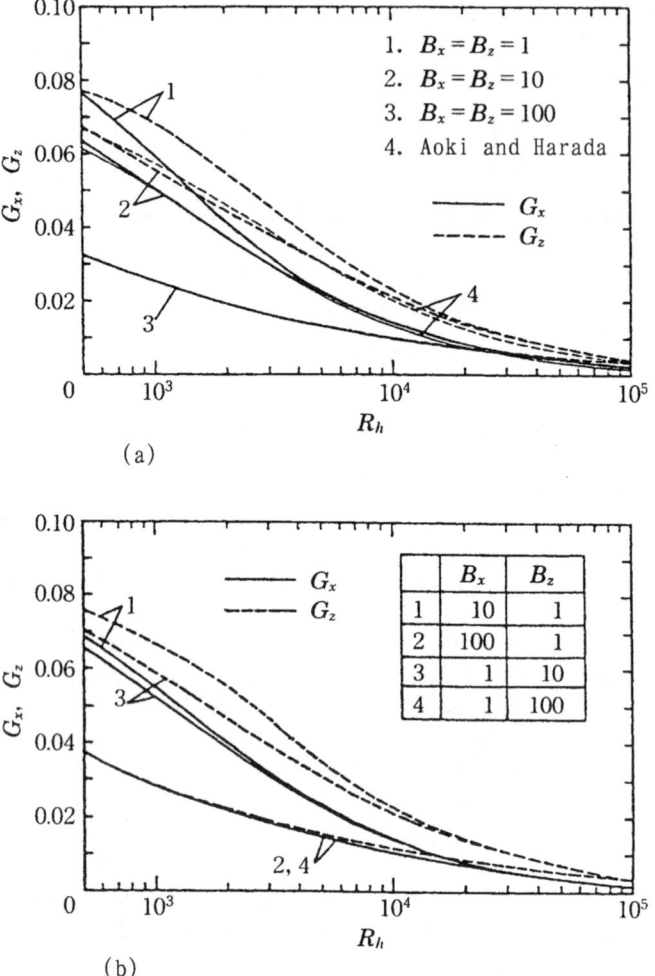

Fig. 9.4a,b. Results of calculation of the turbulent coefficients [33] for equal pressure gradients in the x and z directions (**a**) and for different pressure gradients (**b**)

$$1/G_x = k_x = 12(1 + \alpha_x R_h^{n_x}), \quad 1/G_z = k_z = 12(1 + \alpha_z R_h^{n_z}) \qquad (9.53)$$

where
$$\alpha_x = 0.00116, \quad \alpha_z = 0.00120, \quad n_x = 0.916, \quad n_z = 0.854$$

These expressions were obtained using Eq. 9.33 as a coefficient of modification for the mixing length. R_h is the local Reynolds number. If R_h is sufficiently small, $G_x = G_z = 1/12$ is obtained from the above, and therefore Eqs. 9.51 and 9.52 are in agreement with Reynolds' equation in the case of laminar flow.

9.4 Comparison of Analyses Using the Mixing Length Model with Experiments

In this section, some results of analyses of turbulent fluid film seals based on the mixing length model will be compared with experimental results.

9.4.1 Turbulent Static Characteristics of Fluid Film Seals

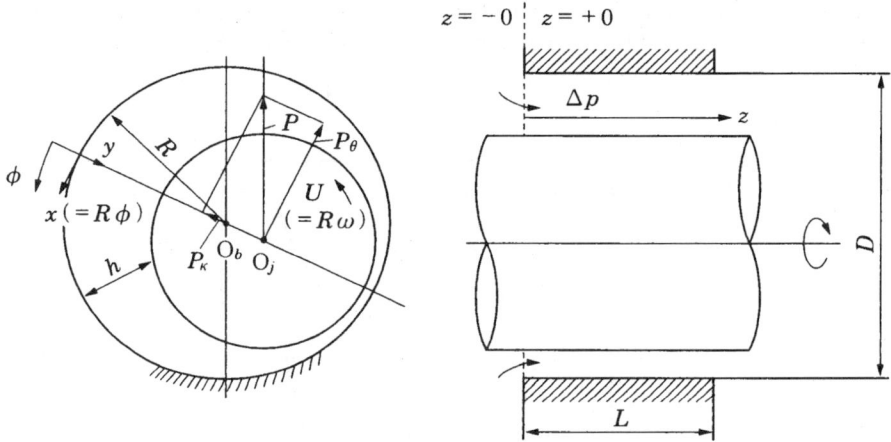

Fig. 9.5. Fluid film seal [35]

The turbulent static characteristics of a fluid film seal as shown in Fig. 9.5 are considered (Kaneko et al. [32] [35]). The turbulent Reynolds' equation (Eq. 9.51 or 9.52) is applied to the film of a seal and is solved by using the finite difference method. In the case of a fluid seal, a big pressure difference exists between the two ends, and a pressure loss takes place at the high pressure end where the liquid flows into the seal clearance. This is taken into consideration in the boundary conditions. The pressure loss is assumed to be expressed as follows when an average inflow velocity in the axial direction is w_m:

$$\Delta p = C_L \frac{\rho w_m^2}{2} \tag{9.54}$$

where C_L is a pressure loss coefficient, which, according to experiments, is given as follows:

$$C_L = -R_0/2900 + 2.57, \quad R_0 = w_m h/\nu \tag{9.55}$$
$$\text{with a proviso that } C_L = \text{constant when } R_0 < 1000. \tag{9.56}$$

212   9 Turbulent Lubrication

R_0 is the local Reynolds' number in the axial direction.

When the journal and seal are eccentric, the average flow velocity w_m changes with the position on the seal circumference, and hence the pressure loss also changes with the position on the circumference. To solve Eq. 9.51 for the pressure distribution, the pressure losses on the seal circumference must be known as a boundary condition, and the pressure distribution must be known beforehand to know the pressure loss because it is a function of inflow velocity. Thus, an iterative calculation is needed.

If the pressure distribution can be found in this way, the load capacity, the journal center loci, and so on of a fluid film seal in a turbulent condition can be obtained. The method of calculation is the same as that for a journal bearing. In the case of a fluid film seal, the pressure difference p_d acting on the seal affects these characteristics.

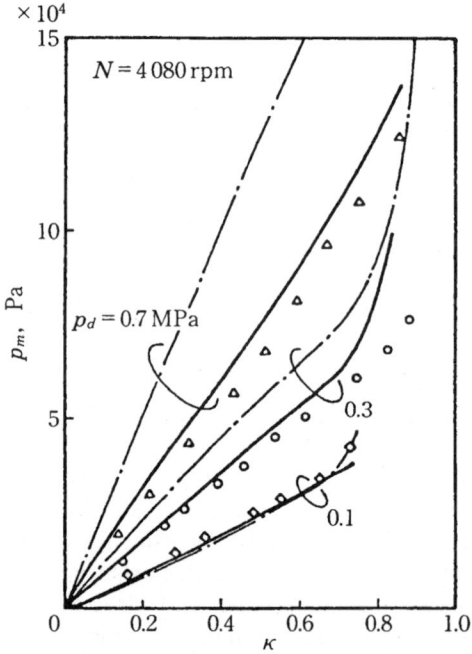

Fig. 9.6. Load capacity and eccentricity ratio in a fluid film seal. *Chained line*, laminar flow; *solid lines*, turbulent flow (theoretical values). *Sysmbols* show experimental values: *diamonds*, pressure difference $p_d = 0.1$ MPa; *circles*, $p_d = 0.3$ MPa; *triangles*, $p_d = 0.7$ MPa [35]

Figure 9.6 shows the relationship between the seal pressure (load capacity per unit area) p_m and the eccentricity ratio κ at a rotating speed of $N = 4080$, with a pressure difference p_d between the two ends of the seal as a parameter. The experimental values are close to the theoretical values in the case of turbulent flow.

9.4 Comparison of Analyses Using the Mixing Length Model with Experiments

It is seen in Fig. 9.6 that the seal pressure p_m increases with an increase in the eccentricity ratio. This is quite natural. On the other hand, it is very interesting to note that the seal pressure p_m increases with the increase in the pressure difference p_d. When a journal is in an eccentric position in a seal, the inflow velocity of the fluid and hence the pressure loss is large at the circumferential position where the seal clearance is large, whereas the pressure loss is small, in contrast, at the position where the seal clearance is small. The difference of the pressure loss yields a static bearing effect in the oil film, and this is added to the oil film force produced by journal rotation. The larger the pressure difference at the two ends of a seal p_d, the larger the static bearing effect and the load capacity (per unit area) p_m will be.

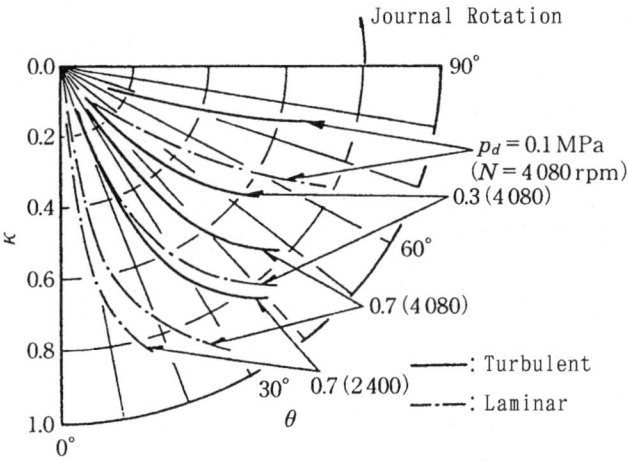

Fig. 9.7. Loci of journal center in a fluid film seal [35]

Figure 9.7 shows the calculated loci of the journal center with pressure difference p_d and journal rotating speed N as parameters. In this case, the loci of the journal center are closer to the vertical line (the direction of loading) for larger p_d, or for smaller N. This is because the static bearing effect due to the pressure difference p_d contributes mainly to the P_κ component of the oil film force but does not contribute to the P_θ component. Consequently, the larger the pressure difference is, the smaller the attitude angle is. When the rotating speed is low, the static bearing effect becomes relatively large, hence the attitude angle becomes small.

9.4.2 Turbulent Dynamic Characteristics of Fluid Film Seals

The elastic and damping coefficients of a turbulent fluid film seal are considered. In this case also, the pressure difference between the two ends of the seal affects these coefficients (Kaneko et al. [32] [36] [37]).

In reference to the coordinate axes x and y of Fig. 9.8, elastic coefficient K_{ij} and damping coefficient C_{ij} are defined as Eq. 9.57 using the oil film forces P_x and P_y, where the positive direction of P_x and P_y are taken in the direction of $-x$ and $-y$. The subscript 0 denotes the static equilibrium position. To obtain P_x and P_y, first solve the turbulent Reynolds' equation (Eq. 9.51) for the pressure, multiply the obtained pressure by $\cos\phi$ and $\sin\phi$, integrate these with respect to ϕ to obtain P_κ and P_θ (see Fig. 9.5), and finally transform these into P_x and P_y.

$$K_{xx} = \left.\frac{\partial P_x}{\partial x}\right|_0, K_{xy} = \left.\frac{\partial P_x}{\partial y}\right|_0, K_{yx} = \left.\frac{\partial P_y}{\partial x}\right|_0, K_{yy} = \left.\frac{\partial P_y}{\partial y}\right|_0$$

$$C_{xx} = \left.\frac{\partial P_x}{\partial \dot{x}}\right|_0, C_{xy} = \left.\frac{\partial P_x}{\partial \dot{y}}\right|_0, C_{yx} = \left.\frac{\partial P_y}{\partial \dot{x}}\right|_0, C_{yy} = \left.\frac{\partial P_y}{\partial \dot{y}}\right|_0 \quad (9.57)$$

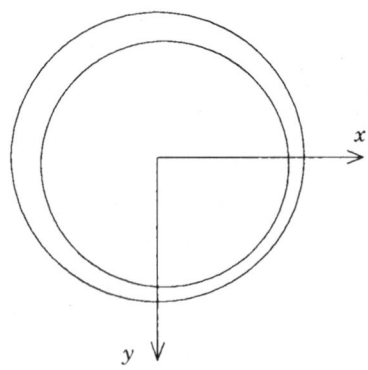

Fig. 9.8. Axes of coordinates (horizontal, vertical)

Figure 9.9 shows three examples of the above-mentioned elastic coefficients and a damping coefficient K_{xx}, K_{xy}, and C_{xx} as functions of eccentricity ratio κ_0 with P_d as a parameter. The figure shows that the larger the pressure difference is, the larger the value of these constants (absolute values) becomes. Further, the value of K_{xx} for laminar flow is larger than that in the case of turbulent flow. The same can be said of K_{yy}, although the data are not shown. The calculation conditions are as follows: $D = 70$ mm, $L = 35$ mm, $c = 0.175$ mm, $N = 4000$ rpm, $\mu = 1.44$ mPa·s, the axial Reynolds' number $R_a = w_m c/v = 767-2540$ (w_m is the axial average velocity), the circumferential Reynolds' number $R_\omega = R\omega c/v = 1418$.

9.5 Turbulent Lubrication Theory Using the k-ε Model

In the case of a journal bearing in which the eccentricity ratio of the journal is large, the pressure gradient in the oil film is large. The mixing length used in the mixing

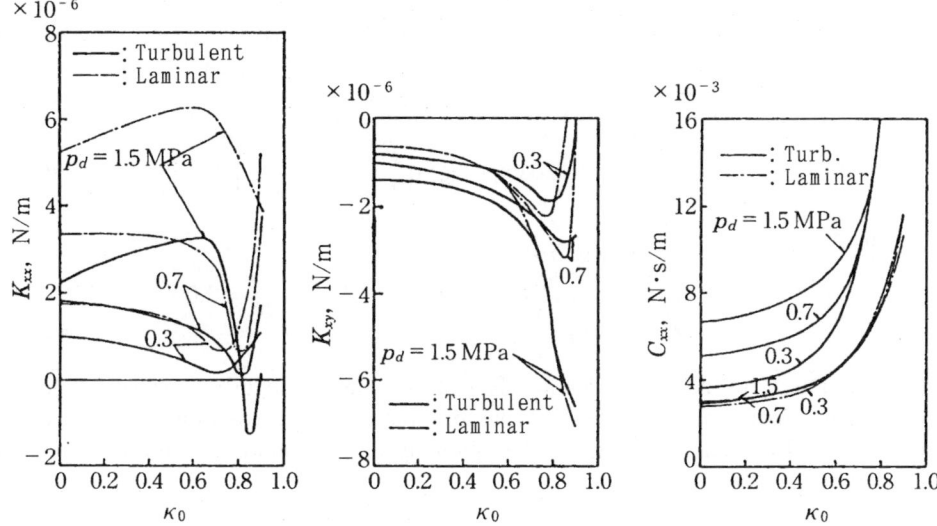

Fig. 9.9. Spring and damping coefficients of a fluid film seal [36]

length model is usually determined experimentally for small pressure gradients, but it changes with pressure gradient. Therefore, it is more reasonable to use the k-ε model, which is less affected by pressure gradient, for analyses of a turbulent bearing with a large eccentricity ratio.

An analysis based on the k-ε model will be described below (Kato and Hori [31]).

9.5.1 Application of the k-ε Model to an Oil Film

In the oil film of a bearing, especially in the neighborhood of the wall surface, the turbulent Reynolds' number R_t is relatively low. The low-Reynolds' number k-ε model, which is suitable in such a case, was introduced by Jones and Launder [21] [22] as stated in the previous section. The transport equations (Eqs. 9.22 and 9.23) for k and ε given by them were improved later by Hassid and Poreh [25]. We use Hassid and Poreh's model here. This model can be applied to cases in which the Toms effect (described later) appears [34].

First, the turbulent energy k and the turbulent loss ε are defined as follows:

$$k = \frac{1}{2}\overline{u_i' u_i'} = \frac{1}{2}\left(\overline{u'^2} + \overline{v'^2} + \overline{w'^2}\right) \tag{9.58}$$

$$\varepsilon = \nu \overline{\frac{\partial u_i'}{\partial x_j}\frac{\partial u_i'}{\partial x_j}} = \nu \overline{\left(\frac{\partial u'}{\partial x}\right)^2} + \cdots + \overline{\left(\frac{\partial w'}{\partial z}\right)^2} \tag{9.59}$$

The following two-dimensional equations, which are extentions of the one-dimensional equations by Hassid and Poreh, will be used as transport equations for k and ε:

$$\frac{Dk}{Dt} = \frac{\partial}{\partial y}\left[\left(v + \frac{v_t}{\sigma_k}\right)\frac{\partial k}{\partial y}\right] - \overline{u'v'}\frac{\partial u}{\partial y} - \overline{v'w'}\frac{\partial w}{\partial y} - \varepsilon - \frac{2vk}{b^2} \quad (9.60)$$

$$\frac{D\varepsilon}{Dt} = \frac{\partial}{\partial y}\left[\left(v + \frac{v_t}{\sigma_\varepsilon}\right)\frac{\partial \varepsilon}{\partial y}\right] - C_{\varepsilon 1}\left(\overline{u'v'}\frac{\partial u}{\partial y} + \overline{v'w'}\frac{\partial w}{\partial y}\right)\frac{\varepsilon}{k}$$

$$-C_{\varepsilon 2}\left[1 - 0.3\exp(-R_t^2)\right]\frac{\varepsilon^2}{k} - 2v\left(\frac{\partial \varepsilon^{1/2}}{\partial y}\right)^2 \quad (9.61)$$

where
$$b = \min(y, h - y)$$
$$\sigma_k = 1, \quad \sigma_\varepsilon = 1.3, \quad C_{\varepsilon 1} = 1.45, \quad C_{\varepsilon 2} = 2.0$$
$$R_t = k^2/(\varepsilon v)$$

In Eqs. 9.60 and 9.61, ε denotes the isotropic part of the turbulent loss and $2vk/b^2$ the anisotropic part. The idea of separating the turbulent loss in this way was proposed by Jones and Launder, and the following simple boundary conditions for ε has thereby become possible:

$$\varepsilon = 0 \quad \text{at the wall surface} \quad (9.62)$$

Further, the boundary conditions for k is assumed to be:

$$k = 0 \quad \text{at the wall surface} \quad (9.63)$$

According to Hassid and Poreh, the turbulent dynamic viscosity coefficient v_t is given as follows by using the above-mentioned k and ε:

$$v_t = C_m \frac{k^2}{\varepsilon}\left[1 - \exp(-A_d R_t)\right] \quad (9.64)$$

However, the following equation, which contains a correction factor C_d, will be used here, based on Laufer's experiments on Couette flows [4]:

$$v_t = C_m \frac{k^2}{\varepsilon}\left[1 - C_d \exp(-A_d R_t)\right] \quad (9.65)$$

where
$$C_m = 0.09, \quad A_d = 1.5 \times 10^{-3}, \quad C_d = 0.95$$

Although a strong Couette flow in the circumferential direction and a weak pressure flow in the axial direction are expected to exist in a bearing, it is assumed here that Eq. 9.65 can be used in both the circumferential and the axial directions, based on the fact that the correction factor $C_d = 0.95$ is close to 1.

9.5.2 Turbulent Reynolds' Equation

The time-average equation of motion of an incompressible fluid containing Reynolds' stress in a two-dimensional case can be given by Eqs. 9.8 and 9.9. In general, it can be written in a tensor expression as follows:

9.5 Turbulent Lubrication Theory Using the k-ε Model

$$\rho\left(\frac{\partial u_i}{\partial t} + u_j\frac{\partial u_i}{\partial x_j}\right) = -\frac{\partial p}{\partial x_i} + \frac{\partial}{\partial x_j}\left(\mu\frac{\partial u_i}{\partial x_j} - \rho\overline{u_i'u_j'}\right) \tag{9.66}$$

(i, j = 1, 2, 3; summation is taken over all values of j)

The equations in rectangular coordinates (x, y, z) can be obtained by the substitution of variables such as $x = x_1$, $y = x_2$, $z = x_3$, $u = u_1$, $v = u_2$, and $w = u_3$, where x, y, and z are the coordinates in the circumferential direction, across the film thickness, and in the axial directions; u and u' (and similar) express the static (time-average) parts of the flow velocity and the fluctuations about it, respectively.

Considering a sufficiently thin lubricating film, let us make the following assumptions.

1. The left-hand side (inertia term) of Eq. 9.66 is negligible.
2. In Eq. 9.66, the derivatives of Reynolds' stresses with respect to x and z can be neglected compared with that with respect to y.
3. The normal components of Reynolds' stress (components $i=j$) can be neglected.
4. $-\rho\overline{u'v'}$ and $-\rho\overline{v'w'}$ can be expressed as follows with a turbulent viscosity coefficient ν_t, which is common to the x and z directions:

$$-\rho\overline{u'v'} = \rho\nu_t\frac{\partial u}{\partial y} \tag{9.67}$$

$$-\rho\overline{v'w'} = \rho\nu_t\frac{\partial w}{\partial y} \tag{9.68}$$

Under these assumptions, a turbulent lubrication equation is derived from Eq. 9.66. First, disregarding the left-hand side of Eq. 9.66 from assumption 1, then substituting Eqs. 9.67 and 9.68 of assumption 4 into this leads to the following equations:

$$\frac{\partial p}{\partial x} = \rho\frac{\partial}{\partial y}\left[(\nu + \nu_t)\frac{\partial u}{\partial y}\right] \tag{9.69}$$

$$\frac{\partial p}{\partial z} = \rho\frac{\partial}{\partial y}\left[(\nu + \nu_t)\frac{\partial w}{\partial y}\right] \tag{9.70}$$

where $\partial p/\partial y = 0$ is omitted. Integrate Eqs. 9.69 and 9.70 twice with respect to y under the boundary conditions

$$u = U_1, \quad w = 0 \quad \text{at} \quad y = 0 \tag{9.71}$$
$$u = w = 0 \quad \text{at} \quad y = h \tag{9.72}$$

to obtain u and w, respectively. Substituting these into the continuity equation

$$\int_0^h \frac{\partial u}{\partial x}dy + \int_0^h \frac{\partial w}{\partial z}dy = 0 \tag{9.73}$$

gives the following turbulent Reynolds' equation, with $G_x = G_z = G$:

$$\frac{\partial}{\partial x}\left(G\frac{\partial p}{\partial x}\right) + \frac{\partial}{\partial z}\left(G\frac{\partial p}{\partial z}\right) = U_1\frac{\partial F}{\partial x} \tag{9.74}$$

where

$$G = \frac{\int_0^h \int_0^y \frac{dy\,dy}{\nu+\nu_t} \int_0^h \frac{y\,dy}{\nu+\nu_t}}{\int_0^h \frac{dy}{\nu+\nu_t}} - \int_0^h \int_0^y \frac{y\,dy\,dy}{\rho(\nu+\nu_t)} \quad (9.75)$$

$$F = h - \frac{\int_0^h \int_0^y \frac{dy\,dy}{\nu+\nu_t}}{\int_0^h \frac{dy}{\nu+\nu_t}} \quad (9.76)$$

If Eq. 9.65 is used, the five equations, Eqs. 9.60, 9.61, 9.69, 9.70, and 9.74 form a closed set of equations with respect to the five unknowns k, ε, u, w, and p. Unknowns such as the fluctuations in the velocity are not included in the equations. Therefore, by solving these five equations simultaneously, the above five unknowns will be obtained. It is assumed here that the left-hand side of Eqs. 9.60 and 9.61 (time variations of k and ε along the streamline) can be disregarded, considering the stationary state:

$$\frac{Dk}{Dt} = 0, \qquad \frac{D\varepsilon}{Dt} = 0$$

Turbulent lubrication problems can thus be solved.

Approximate solutions are possible when the left-hand side (inertia term) of Eq. 9.66 cannot be disregarded [28].

9.6 Comparison of Analyses Using the k-ε Model with Experiments

In this section, some examples of comparisons of theoretical analyses of a turbulent bearing by the k-ε model and experiments will be shown (Kato and Hori [31]). In theoretical calculations, Eqs. 9.60, 9.61, and 9.74 are solved simultaneously under the boundary conditions given in Eqs. 9.71, 9.72, 9.62, 9.63 and the following boundary condition concerning pressure:

$$p = 0 \quad \text{at} \quad \theta = 0, \ \pi \ \text{and at the bearing ends.} \quad (9.77)$$

The procedure for numerical calculations is as follows. Assume suitable initial profiles of k and ε, calculate G and F of Eqs. 9.75 and 9.76 and then obtain the pressure distribution by applying the finite element method to the turbulent lubrication equation (Eq. 9.74). Next, calculate the flow velocity distributions u and w from the pressure distribution by using Eqs. 9.69 and 9.70 and the boundary conditions Eqs. 9.71 and 9.72, then obtain k and ε from the above velocity distributions, Eq. (9.60) and Eq. (9.61). Using these results, calculate G and F again, and then obtain

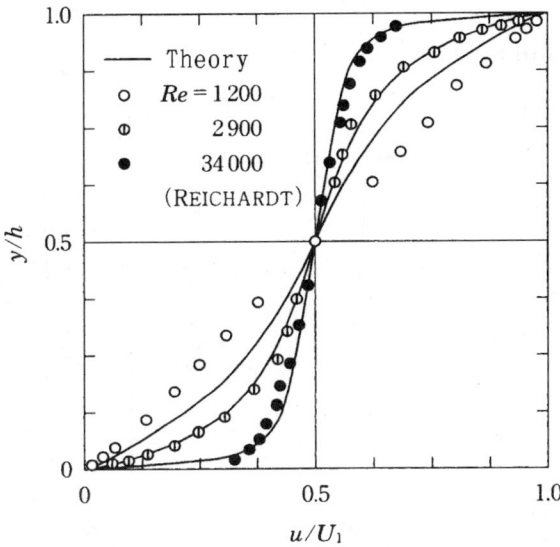

Fig. 9.10. Average velocity distribution of Couette flow [31]

the pressure distribution again in the same way as in the beginning of the procedure. This calculation will be repeated until the calculated pressure distribution converges within a small error.

As an example, consider a journal with a diameter of 150 mm rotating in a bearing with an inner diameter of 152 mm and a length of 150 mm over the speed range 100–5000 rpm. Let the coefficient of dynamic viscosity of the lubricating oil be 9.4×10^{-6} m^2/s (30°C).

Comparisons of the theoretical and the experimental results, both under the above conditions unless otherwise stated, will be shown below. Lubricating oil is supplied at a rate of 12 l/min in the experiments.

Figure 9.10 shows the theoretical and the experimental results of the average velocity distribution u for Couette flow, the experiment by Reichardt being used in this case [6]. It is seen in Fig. 9.10 that although the theoretical and the experimental results are different for a Reynolds' number of $Re = 1200$, they are in good agreement for $Re = 2900$ and $Re = 34\,000$. This shows that the k-ε model is better suited to analysis in the turbulent region well above the laminar-to-turbulent transition region.

Figure 9.11a,b shows the theoretical and experimental results for the pressure distribution in a finite width bearing. Figures 9.11a,ba,b are the nondimensional pressure distribution $\bar{p}$ in the circumferential direction at the center of bearing width for $Re = 2000$ and 8000, respectively, the parameter being the eccentricity ratios as shown. The theoretical and the experimental values are generally in good agreement. However, for $Re = 8000$ and an eccentricity ratio of 0.8, the experimental pressure

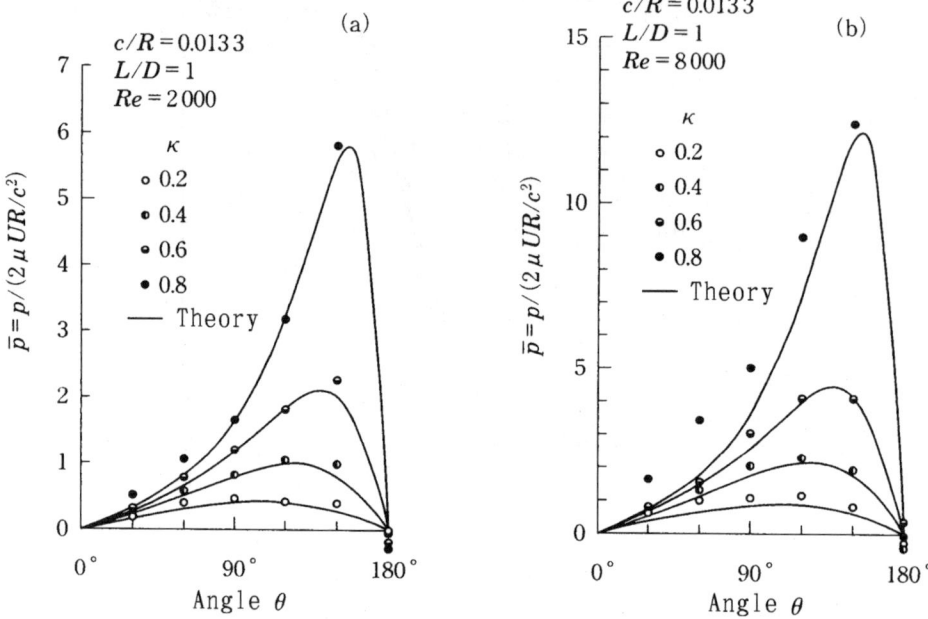

Fig. 9.11a,b. Theoretical (*symbols*) and experimental (*lines*) values of the nondimensional pressure distribution in a finite width bearing (1) [31]

is higher than the theoretical pressure in the range $0°-90°$. This may be attributed to the influence of the inertia of the fluid.

In Fig. 9.12, the nondimensional load capacity $\bar{P} = P/(2\mu U R^2 L/c^2)$ (the left scale) is plotted against eccentricity ratio κ for $Re = 2000$ and 5000. The reciprocal of Sommerfeld's number S^{-1} is also shown in the figure (the right scale). The relation $S^{-1} = 2\pi\bar{P}$ holds between the two axes. The theoretical and experimental results of load capacity or Sommerfeld's number are in good agreement even when the eccentricity ratio exceeds 0.95. This shows that the lubrication theory based on the k-ε model can be applied to bearings with very high eccentricity ratios.

Figure 9.13a,b shows the nondimensional pressure distributions $\bar{p}$ for the cases shown in Fig. 9.12. The parameter for the curves is the nondimensional load capacity $\bar{P}$ of Fig. 9.12. The theoretical and experimental values for the pressure distribution are in good agreement even in the case of high eccentricity ratios over 0.9. This also shows that the turbulent lubrication theory based on the k-ε model can be used for bearings of high eccentricity ratio. This is because the k-ε model is valid for large pressure gradients.

Figure 9.14 shows the theoretically obtained loci of the journal center with those obtained experimentally by Wada and Hashimoto [27]. The figure shows that

9.6 Comparison of Analyses Using the k-ε Model with Experiments   221

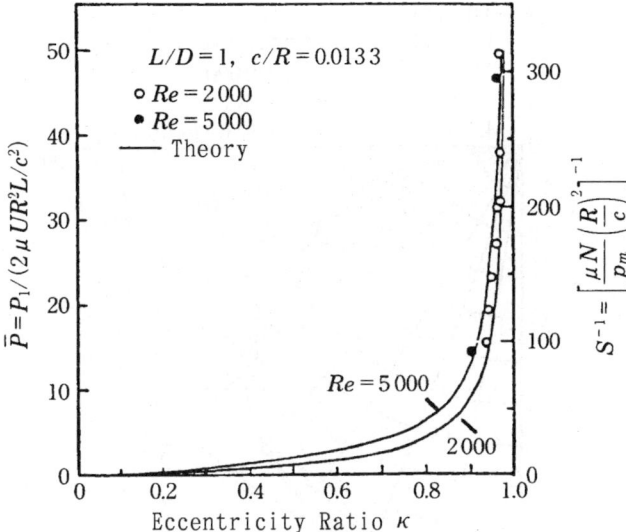

Fig. 9.12. Nondimensional load capacity and Sommerfeld's reciprocal versus eccentricity ratio [31]

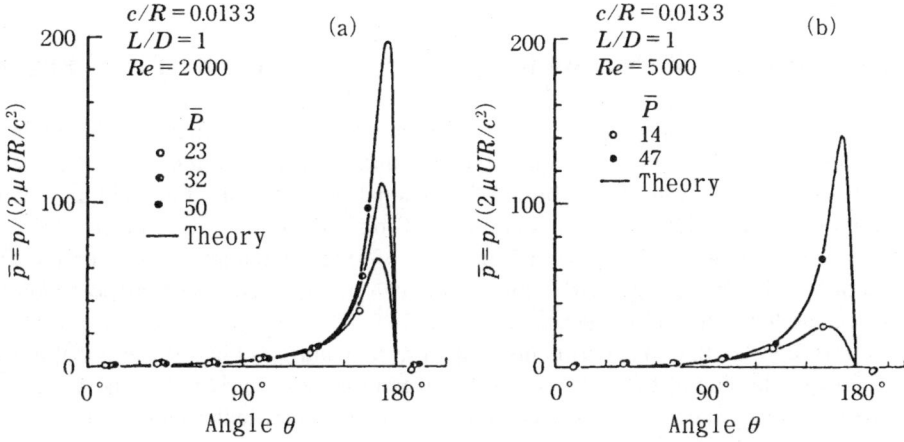

Fig. 9.13a,b. Nondimensional pressure distribution in a finite width bearing (2) for the two cases analyzed in Fig. 9.12 [31]

changes in Reynolds' number do not very much affect the shape of the loci of the journal center.

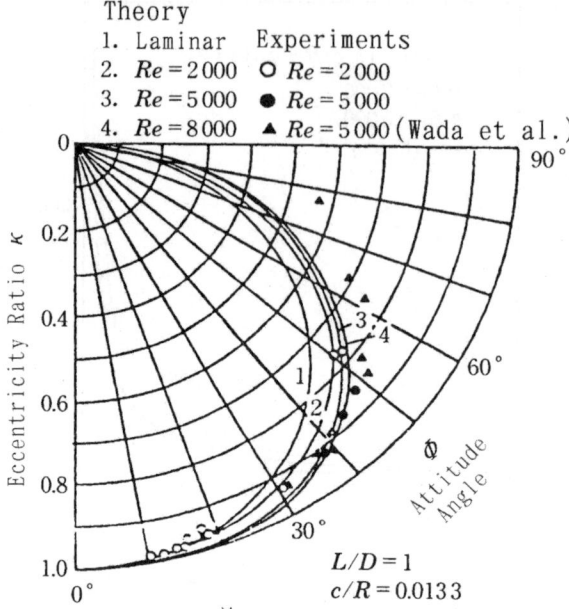

Fig. 9.14. Locus of the journal center [31]

9.7 Reduction of Friction in a Turbulent Bearing by Toms' Effect

Large shear stress and large heat generation in a fluid film are major problems in a turbulent bearing. In pipe flow, on the other hand, the marked reduction in turbulent flow resistance of water by addition of very small amounts of a kind of long-chain high molecular weight polymer is known as Toms' effect [2] [19]. Since it is known that Toms' effect is more powerful in pipes of smaller diameter, considerable reduction in the friction of a turbulent bearing can be expected if this phenomenon is applied to the thin film of a bearing.

It is reported that in experiments using a water solution of hydroxyethylcellulose or polyacrylamide of the order of ppm between two concentric cylinders (diameter of the inner cylinder 10 mm, clearance 1 mm), the turbulent friction for Couette flow was actually reduced, and that, in the case of two eccentric cylinders, an even larger reduction in friction was observed for larger eccentricity ratios [23]. It has also been reported that a similar reduction in friction was observed in experiments using a water solution of polyethyleneoxide in a far smaller clearance, but the effect was lost after only 20 circulations of the solution [24].

In this section, results of experiments investigating Tom's effect under the conditions close to those of an actual bearing (Fukayama et al. [29]) and comparisons of them with calculated results based on the k-ε model (Kato and Hori [34]) are shown.

9.7 Reduction of Friction in a Turbulent Bearing by Toms' Effect

The following conditions were used: bearing diameter 210 mm, bearing length 200 mm, bearing clearance 0.202 mm, and, as the lubricating fluid, pure water and a water solution of polyacrylamide (PAA, molecular weight 2.3×10^6).

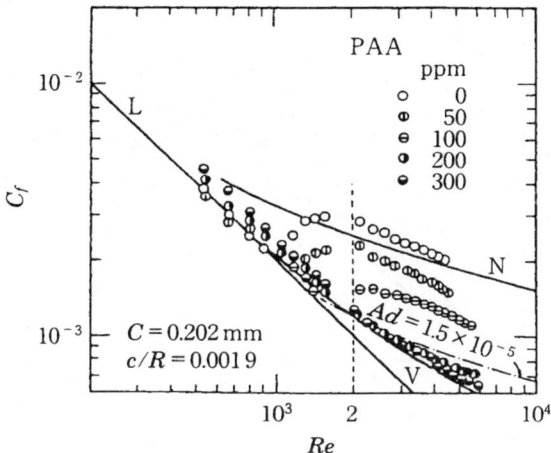

Fig. 9.15. Coefficient of friction of a bearing using various concentrations of polyacrylamide (*PAA*) in water as the lubricating fluid [29] [34]

Figure 9.15 shows the experimentally obtained relations between Reynolds' number $Re = Uc/\nu$ and the coefficient of friction of the bearing $C_f = 2\tau_w/(\rho U^2)$ (τ_ω is the shear stress at the wall surface). The straight line L corresponds to laminar flow and the curve N to turbulent flow of pure water. The curve V corresponds to Virk's maximum drag reduction which was obtained for a 300 ppm solution of PAA in water. Comparison of curve V and the calculated results based on the k-ε model gives the value of $A_d = 1.5 \times 10^{-5}$ for the constant A_d in Eq. 9.64 in the case of maximum drag reduction.

Figure 9.16 shows the experimental value of the nondimensional pressure distribution $\overline{p}$ and the theoretical distribution based on the k-ε model. Curves 1 and 2 in Fig. 9.16 show the pressure distribution for pure water and a solution of PAA in water, respectively, calculated using the k-ε model. They are in good agreement with the experimental results shown by small circles. As the values of constant A_d in Eq. 9.64, $A_d = 1.5 \times 10^{-3}$ was used for pure water and $A_d = 1.5 \times 10^{-5}$, as obtained above, was used for the PAA solution. Thus, the k-ε model can explain well the pressure distribution in the bearing for both pure water and PAA solutions. While the mixing length model gives good agreement in the case of pure water, it is known to estimate the pressure distribution in the case of PAA solutions 15%–20% too low.

According to Fig. 9.16, which is for $Re = 2000$, the drop in pressure (decrease in load capacity) due to the addition of PAA is not very large. On the other hand, Fig. 9.15 shows for $Re = 2000$ (shown by the dashed line) that the fall in the frictional

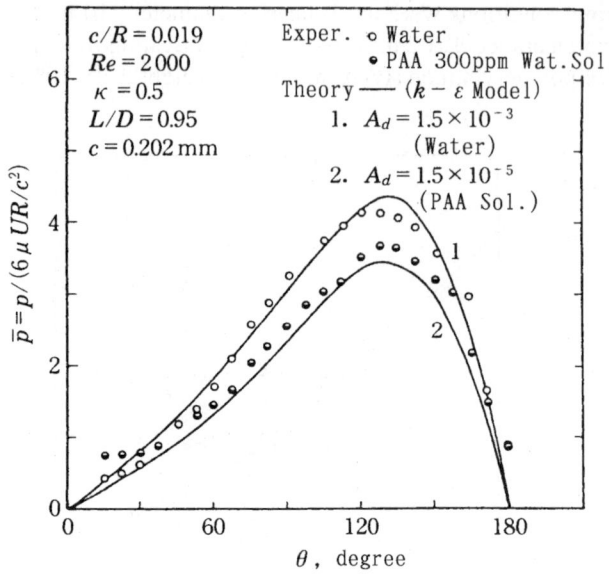

Fig. 9.16. Nondimensional pressure distribution for water and a PAA solution [34]

resistance due to the addition of PAA is very large. In short, although the reduction in load capacity due to the addition of PAA is not very large, the reduction in the frictional resistance is very large, which is very favorable for bearing design.

Thus, Toms' effect is effective in a turbulent bearing and can be analyzed using the k-ε model.

9.8 Taylor Vortices in a Journal Bearing

The fluid in the clearance between two concentric cylinders becomes unstable when the rotating speed of the inner cylinder reaches a certain critical speed, the outer cylinder being at rest, and beyond that speed the fluid begins to exhibit a series of regularly spaced rings of vortices around the inner cylinder. Those are called **Taylor vortices** and have been known as beautiful and interesting phenomena for many years [1]. The following is known about Taylor vortices:

1. The critical speed for the onset of Taylor vortices is given by $U_i h/\nu > 4.13 \sqrt{R_i/h}$, where R_i is the radius of the inner cylinder, U_i is the surface velocity of the inner cylinder, h is the clearance between the two cylinders, and ν is the coefficient of kinetic viscosity.
2. The pitch of the Taylor vortices in the axial direction (width of a vortex) is approximately equal to the clearance of the two cylinders.

Further increases in the speed of the inner cylinder finally lead to complete turbulence of the fluid. In other words, the Taylor vortices appear in the laminar–turbulent transitional region.

The Taylor vortices can develop also in the oil film of a journal bearing, and are considered to affect the bearing characteristics in the laminar–turbulent transitional region. However, the situation in a journal bearing is different from that of usual Taylor vortices in that the journal and bearing are generally not concentric. Although many experimental and theoretical studies have been done since Taylor on the stability of a fluid between two concentric cylinders, not very much has been done on that of a fluid between two eccentric cylinders. Some relevant papers are listed in the References [8] [13] [18].

In this section, results of visualization experiments on Taylor vortices between two eccentric cylinders will be described (Kato and Hori [30]). The specifications of the experimental apparatus are: diameter of the inner cylinder (journal) 150 mm, inner diameter of the outer transparent cylinder (bearing) 152 mm, length of the test section of the journal 152 mm, and the coefficient of kinetic viscosity of the fluid (base oil) 9.4×10^{-6} m^2/s (at 30 °C). Aluminum flakes were suspended in the fluid for the purpose of visualization.

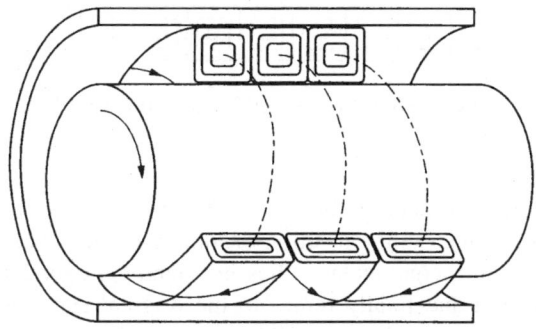

Fig. 9.17. Taylor vortices between two eccentric cylinders [30]

The following were observed with increases in the rotational speed of the inner cylinder:

1. At a certain critical rotating speed, the flow which was laminar until then becomes unstable in the neighborhood of the maximum clearance and the Taylor vortices develop there. The pitch of the vortices is nearly equal to the clearance at this position (the maximum clearance in this case). The Taylor vortices do not exist yet in the domain of smaller clearance.
2. With further increases in the rotational speed of the inner cylinder, the domain of instability spreads toward areas of smaller clearance, the Taylor vortices extend into the region of smaller clearance, and finally form complete rings of

Taylor vortices around the inner cylinder. Inspection of a vortex from the maximum clearance position toward the minimum clearance position reveals that the vortex flattens gradually and the pitch widens gradually. Figure 9.17 shows the situation. The vortex is flattest and widest at the position of minimum clearance.

3. However, it is quite unnatural that, when the eccentricity ratio is large, the Taylor vortices are very flat and wide in the area of minimum clearance, because the pitch of Taylor vortices tends to be nearly equal to the clearance of the two cylinders (the minimum clearance in this case). Therefore, a vortex that has become too flat branches into more stable, thinner vortices, the pitch of which is nearly equal to the clearance (local stability). This situation is shown in Fig. 9.18. An assumption that the fluid of two adjoining vortices moves in the same direction at the border leads to the conclusion that one vortex branches into at least three vortices, and in general into an odd number of vortices, as shown in the figure.

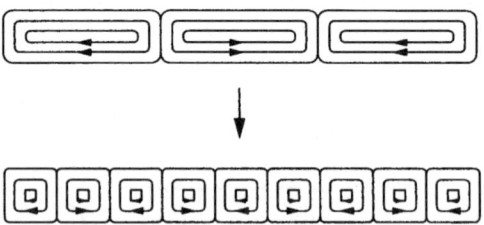

Fig. 9.18. Branching of Taylor vortices [30]

4. The Taylor vortices branch in this way and result in vortices of minimum pitch and maximum number at the position of minimum clearance. The vortices that thus branched will merge again as they go round the cylinder to the maximum clearance position and finally return to the original state. The branching and merging may occur irregularly, resulting in somewhat disturbed vortices. Based on the above observations, a local stability theory of the flow between two eccentric cylinders and a hydrodynamic lubrication theory taking the Taylor vortices into consideration were proposed [30].

References

1. G.I. Taylor, "Stability of a Viscous Liquid Contained Between Two Rotating Cylinders", *Phil. Trans.*, A223, 1923, pp. 289-343.; *Proc. Roy. Soc.*(London), A151, 1935, p. 494; *ibid*, A157, 1936, p. 546 and p. 565.
2. B.A. Toms, "Some Observations on the Flow of Linear Polymer Solutions Through Straight Tubes at Large Reynolds Numbers", *Proc. of 1st International Congress on Rheology*, North Holland Publishing Co., Amsterdam, Vol. 2, 1949, pp. 135- .
3. D.F. Wilcock, "Turbulence in High-Speed Journal Bearings", *Trans. ASME*, Vol. 72, August 1950, pp. 825-834.

4. J. Laufer, "The Structure of Turbulence in Fully Developed Pipe Flow", NACA Tech. Report 1174, Washington, D.C., 1953.
5. E.R. van Driest, "On Turbulent Flow Near a Wall", *Journal of Aeronautical Science*, Vol. 23, No. 11, November 1956, pp. 1007-1011.
6. H. Reichardt, "Über die Geschwindigkeitsverteilung in einer geradlinigen turblenten Couettestroemung", *ZAMM-Sonderheft*, Vol. 36 (1956), pp. S 26 - S 29.
7. V.N. Constantinescu, "On Turbulent Lubrication", *Proceedings of the Institution of Mechanical Engineers*, Vol. 173, 1959, pp. 881-900.
8. R.C. DiPrima, "A Note on the Stability of Flow in Loaded Journal Bearings", *ASLE Trans.*, Vol. 6, 1963, pp. 249-253.
9. Chung-Wah Ng and C.H.T. Pan, "A Linearized Turbulent Lubrication Theory", *Journal of Basic Engineering, Trans. ASME, Series D*, Vol. 87, September 1965, pp. 675-688.
10. I. Tani, "Nagare-Gaku (Study of Flow)" (in Japanese), *Iwanami Zensho Series, Iwanami Shoten*, Tokyo, 1967.
11. F.K. Orcut and E.B. Arwas, "The Steady-State and Dynamic Characteristics of a Full Circular Bearing and a Partial Arc Bearing in the Laminar and Turbulent Flow Regimes", *Journal of Lubrication Technology, Trans. ASME, Series F*, Vol. 89, April 1967, pp. 143-153.
12. H.G. Elrod, Jr. and C.W. Ng, "A Theory of Turbulent Fluid Films and its Application to Bearings", *Journal of Lubrication Technology, Trans. ASME, Series F*, Vol. 89, No. 3, 1967, pp. 346-362.
13. J.H. Vohr, "An Experimental Study of Taylor Vortices and Turbulence in Flow Between Eccentric Rotating Cylinders", *Journal of Lubrication Technology, Trans. ASME, Series F*, Vol. 90, No. 1, 1968, pp. 285-296.
14. M. Shirakura and H. Ohashi, "Fluid Mechanics" (in Japanese), *Corona Publishing Co., Ltd.*, Tokyo, 1969.
15. C.M. Taylor, "Turbulent Lubrication Theory Applied to Fluid Film Bearing Design", Paper 6, *Proceedings of the Institution of Mechanical Engineers*, Vol. 184, 1969-70, Pt. 3L, pp. 40-47.
16. E.R. Booser, A. Missana and F.D. Ryan, "Performance of Large Steam Turbine Journal Bearings", *ASLE Transactions*, Vol 13, 1970, pp. 262-268.
17. H. Aoki and M. Harada, "Turbulent Lubrication Theory for Full Journal Bearings" (in Japanese), *Journal of Japan Society of Lubrication Engineers*, Vol. 16, No. 5, May 1971, pp. 348-356.
18. P. Castle, F.R. Mobbs and P.H. Markho, "Visual Observations and Torque Measurements in the Taylor Vortex Regime Between Eccentric Rotating Cylinders", *Journal of Lubrication Technology, Trans. ASME, Series F*, Vol. 93, No. 1, 1971, pp. 121-128.
19. J.W. Hoyt, "The Effect of Additives on Fluid Friction", A Lecture, *Journal of Basic Engineering, Trans. ASME, Series D*, Vol. 94, June 1972, pp. 258-285.
20. B.E. Launder and D.B. Spalding, "Lectures in Mathematical Models of Turbulence", Academic Press, London, 1972
21. W.P. Jones and B.E. Launder, "The Prediction of Laminarization with a Two-Equation Model of Turbulence", *International Journal of Heat Mass Transfer*, Vol. 15, 1972, pp. 301-313.
22. W.P. Jones and B.E. Launder, "The Calculation of Low-Reynolds-Number Phenomena with a Two-Equation Model of Turbulence", *International Journal of Heat Mass Transfer*, Vol. 16, 1973, pp. 1119-1130.
23. H. Aoki, T. Kusaga, H. Tada, H. Sasaki, "A Study on Drag Reduction of Dilute Polymer Solutions in Annular Flow between Rotating Cylinders", *Colloques Internationaux du C.N.R.S.*, Vol. 223, 1975, pp. 345-348.

24. L.G. Hampson and H. Naylor, "Friction Reduction in Journal Bearing by High Molecular Weight Polymers", *Proc. the 2nd Leads-Lyon Symposium on Tribology*, Mechanical Engineering Publications Ltd., London, September 1975, pp. 70-72.
25. S. Hassid and M. Poreh, "A Turbulent Energy Dissipation Model for Flows With Drag Reduction", *Journal of Fluids Engineering, Trans. ASME*, Vol. 100, No. 1, March 1978, pp. 107-112.
26. S. Wada and H. Hashimoto, "Turbulent Lubrication Theory Using the Frictional Law (First Report, Derivation of Turbulent Coefficient and Lubrication Equation)" (in Japanese), *Trans. JSME*, Vol. 44, No. 382, June 1978, pp. 2140-2148.
27. S. Wada and H. Hashimoto, "Ditto (Second Report, Its Application to Journal Bearing)" (in Japanese), *Trans. JSME*, Vol. 44, No. 382, June 1978, pp. 2149-2156.
28. R.E. Hinton and J.B. Roberts, "The Characteristics of a Statically Loaded Journal Bearing with Superlaminar Flow", *Journal of Mechanical Engineering Science, IMechE*, London, Vol. 22, No. 2, 1980, pp. 79-94.
29. H. Fukayama, M. Tanaka and Y. Hori, "Friction Reduction in Turbulent Journal Bearings by Highpolymers", *Journal of Lubrication Technology, Trans. ASME, Series F*, Vol. 102, No. 4, October 1980, pp. 439-444.
30. T. Kato and Y. Hori, "Taylor Vortices in a Journal Bearing" (in Japanese), *Trans. JSME*, Vol. 49, No. 445, September 1983, pp. 1510-1520.
31. T. Kato and Y. Hori, "Turbulent Lubrication Theory Using k-ε Model for Journal Bearings" (in Japanese), *Journal of Japan Society of Lubrication Engineers*, Vol. 28, No. 12, December 1983, pp. 907-914.
32. S. Kaneko, Y. Hori and M. Tanaka, "Static and Dyanmic Characteristics of Annular Plain Seals", *Proc. of Third International Conference on Vibrations in Rotating Machinery*, IMechE, Univ. of York, England, September 11-13, 1984, pp. 205-214.
33. Y. Hori, H. Fukayama, M. Tanaka, T. Kato and S. Kaneko. "Turbulent LubricationTheory for Annular Plain Seals" (In Japanese), *Journal of Japan Society of Lubrication Engineers*, Vol. 30, No. 6, June 1985, pp. 430-437.
34. T. Kato and Y. Hori, "Pressure Distributions in a Journal Bearing Lubricated by Drag Reducing Liquids under Turbulent Conditions", *Proceedings JSLE International Tribology Conference*, July 8-10, 1985, Tokyo, Japan, pp. 571-576.
35. S. Kaneko, Y. Hori, T. Kato and M. Tanaka, "Static Characteristics of Annular Plain Seals in the Turbulent Regime" (in Japanese), *Journal of Japan Society of Lubrication Engineers*, Vol. 31, No. 7, July 1986, pp. 493-500.
36. S. Kaneko, Y. Hori and M. Tanaka, "Dynamic Characteristics of Annular Plain Seals in the Turbulent Regime (First Report, Theoretical Analysis)" (in Japanese), *Journal of Japan Society of Lubrication Engineers*, Vol. 31, No. 9, September 1986, pp. 650-657.
37. S. Kaneko, Y. Hori and M. Tanaka, "Dynamic Characteristics of Annular Plain Seals in the Turbulent Regime (Second Report, Experimental Analysis)" (in Japanese), *Journal of Japan Society of Lubrication Engineers*, Vol. 32, No. 2, February 1987, pp. 141-147.
38. H.K. Myong and N. Kasagi, "A New Proposal for a k-ε Turbulence Model and Its Evaluation (First Report, Development of the Model)" (in Japanese), *Trans. JSME*, C, Vol. 54, No. 507, November 1988, pp. 3003-3009.
39. C. Arakawa, "Computational Fluid Dynamics for Engineering" (in Japanese), *University of Tokyo Press*, Tokyo, 1994.
40. T. Kajishima, "Numerical Simulation of Turbulent Flows" (in Japanese), Yokendo Ltd., Tokyo, 1999.

Index

alignment of bearings   89
Amonton's law   5
animal joint   137
approximate nonlinear analysis   98

bearing number   2
boundary condition
  - of oil film   27
  Gümbel's -   28, 37, 42
  half Sommerfeld's -   29
  Reynolds' -   29
  separation -   29
  Sommerfeld's -   28, 31
  Swift-Stieber's -   29
boundary lubrication   4

cavitation   146
chaos   112
circular bearing   23
circular journal bearings   25
column model   153
constant-strain-rate modulus   154
Coulomb's law   5
critical speed   63
cylindrical coordinates   55

deformation of a pad   58
dissipation energy   170
dry friction   4
dynamic oil film force   71
dynamic oil film pressure   68

energy equation   166, 168
energy loss   1

finite element method   122
finite length (journal) bearing   27, 43
finite length plane pad bearing   54
floating bush bearing   24, 102, 113
fluid film seal   209, 211
foil bearing   119
foil disk   131
friction   1

generator rotor   87

half-speed whirl   65
heat generation   161
Hermann's variational method   150
Holm, R.   6
hydrodynamic bearing   23
hydrodynamic lubrication   3, 6, 9
hydrostatic bearing   23

infinitely long (journal) bearings   29
infinitely long bearing   31
infinitely long plane pad bearing   48
isoviscous anlysis   172

Jost Committee   2
journal bearing   6, 23
journal bearing
  - attitude angle   25
  - clearance circle   25
  - eccentricity ratio   25
  - frictional moment   35, 40, 43
  - infinite length approximation   27
  - load capacity   35
  - oil film constant   75
  - oil film damping constants   75

- oil film force 32, 38, 42
- oil film pressure 29, 31, 37, 41
- oil film spring constant 75
- oil film thickness 26
- radial clearance 25
- shape of oil film 26
- short bearing approximation 27, 41

k-ε model 203, 214
Kármán constant 202
Kingsbury bearing 48
Knudsen number 59, 131

leakage of lubricating oil 43
Leonardo da Vinci 5
limit cycle 98
locus of journal center 33, 39, 42
lubricant 1
lubrication 1
 various forms of - 2

magnetic disk memory device 7, 59
magnetic head 7
magnetic tape memory storage 120, 130
mean free path 59, 130
Michell bearing 48
mixed lubrication 4
mixing length model 201, 204
moiré method 132, 158
moving surface 22
multi-arc bearing 23
multibearing system 89

nonlinear stability 94

oil film rupture 28
oil whip 63, 64
 - hysteresis 64, 84
 - inertia effect 64
 - influence of an earthquake 92
 - preventing method 113
 - theory 67
 secondary - 87
oil whirl 65
Okazaki's method 69

parametric excitation 63
partial bearing 23
Pertrov's law 36
plane pad bearing 48

- center of pressure 52
- frictional force 53
- load capacity 51
- pivot position 53
- pressure distribution 49
porous bearing 109
pressure spike 120

read/write element 59
Reynolds' equation 11, 17
 generalized- 163, 165
Reynolds' stress 199
Reynolds' theory 11
Reynolds, O. 9
Routh-Hurwitz criterion 79

sector pad bearing 55
seizure 1
self-excited vibrations
 - due to oil film action 63
 - due to internal damping 63
 flow-induced - 63
short bearing 41
side leakage factor 54
sinusoidal squeeze 144, 145, 149
skeletal joint 7
sliding bearing 23
slip flow 59, 131
small-end bearing 137
solid friction 4
Sommerfeld transform 30
Sommerfeld's number 34
squeeze effect 7, 18, 137
squeeze film 137
stability chart 80, 81
stability limit 76
stationary surface 21
stretch effect 18
Stribeck diagram 3

Taylor vortex 224
temperature analysis
 - of a circular journal bearing 185
 - tilting pad thrust bearing 172
temperature rise 161
thermohydrodynamic lubrication 162
THL 162
three arc bearing 23, 106, 113
 - offset factor 107

- preload factor   107
thrust bearing   23, 47
tilting pad bearing   23, 48, 113
time-average equation of motion   200
Toms' effect   222
Tower, B.   9
transfer matrix   91
transition temperature
  of oil film   161
tribology   2
  meaning of -   7
true contact area   5
turbulence model   201

turbulent flow energy   203
turbulent flow loss   203
turbulent lubrication   197
turbulent lubrication theory
  - k-ε model   215
  - mixing length model   204
turbulent shear stress   199
turbulent viscosity coefficient   202, 203
two arc bearing   23, 113

viscoelastic model   153

wear   1
wedge effect   7, 18